# GOLD!
## Where to go & how to find it

In *Gold: Where to Go, How to Find It*, author Richard Shears provides the answers to everything the layman ever wanted to know about gold and gold-prospecting. This immensely practical book not only explains why the 'new gold rush' has happened and where and how to look for the precious metal, but also the differences between the new and old gold rushes, with historical detail and advice from old-timers.

Before he was lured to the Australian goldfields, the closest Richard came to any impressive rocks were the granite tors of Dartmoor, England, where he spent much of his childhood. Later he became a journalist, and worked as a foreign correspondent for a Fleet Street newspaper. In 1974 he and his Australian wife, Sally Wilkins, bought an old van and drove it across Asia to Australia. They now live in Melbourne with a stray cat called Ferdinand.

GARRETT

# GOLD!
## Where to go & how to find it

Richard Shears

M

It would be impractical to list everybody who supplied information for this book and who gave valuable advice. They range from old-timers in the hills (who for reasons best known to themselves insisted on remaining anonymous!), to the owners of modern prospecting outlets, to geologists. Special thanks, however, are due to the various departments of mines and minerals in each state; and, in Victoria, to Ron Mileham of Distribeaut, Knoxfield, Collin Prigoone and John Lorimer of Old Gold Mines, Dandenong, and Liz and Leigh Moore of Honor Bright Pty Ltd, East Hawthorn.

Grateful thanks to the Rex Nan Kivell Collection, National Library of Australia, and to the La Trobe Library, Melbourne for permission to use illustrations from their collections.



Created and produced by
Mead & Beckett Publishing
139 Macquarie Street, Sydney

First published 1980 by
THE MACMILLAN COMPANY OF AUSTRALIA PTY LTD
107 Moray Street, South Melbourne 3205
6 George Place, Artarmon 2064

Associated companies in London and Basingstoke, England
New York Dublin Johannesburg Dehli

National Library of Australia cataloguing in publication data
Shears, Richard, 1944-
Gold! Where to go and how to find it.
ISBN 0 333 29933 7

1. Prospecting — Australia. 2. Gold mines and mining — Australia. I. Title.

622'.1841'0994

Design by Barbara Beckett
Edited by Edink
Photoset in Bembo by Meredith Trade Lino, Melbourne
Printed in Hong Kong

# Contents

# 1 The New Gold Rush: Why It Happened

Until Australia entered the 1980s decade, gold had begun to be taken for granted. We hardly noticed the watch band as we checked the time. Admiring eyes stared at the opal and not the chain that held it. Some people crunched food without a thought for the valuable fillings in their teeth. The rich signed their cheques with gold-plated pens and businessmen flicked gold lighters under the cigars of clients. Gold was all around us, as common, it seemed, as brass. Modern scientists had even succeeded where the determined medieval alchemists had failed, and learned how to make it (although they found the cost of the atomic process was prohibitive). If gold was man's most precious metal, there was a great deal of it about.

And then an extraordinary thing happened. Gold fever came to Australia, bringing about a rush to the mining areas that was surpassed only by the fervour of those desperate prospectors of the 1800s. Businessmen were actually driving to the old diggings in their big expense-account cars in order to walk in the cart and wheelbarrow tracks of the old-timers. Some, including a Perth lawyer, even gave up their jobs and professions to search for gold. Trinket boxes were opened in thousands of homes and searching fingers sifted through jewellery that had been ignored for years. Armed guards had to be called in to control the crowds at a Sydney gold refinery. Sales of metal-detectors soared and the rivers came alive with the swishing of pans.

What brought about this incredible craving for gold? In simple terms, modern man had once again come to recognise its value, a pattern that is repeated whenever paper money weakens on the world market. This time, though, gold had become far more valuable than in any peacetime period.

It started its unbelievable climb as the old decade began to fade out. The move was sparked by a seemingly unrelated event — the announcement of President Jimmy Carter's energy policy. Concerned about spiralling oil prices, the President said he wanted to

**When the price of gold increased, people went out to look for their own. This is prospecting, 1980 style.**

invest more than $U.S.88 000 million on alternative fuel research. The world's money-dealers didn't want to hear that. The President's energy policy, concentrating on an overall plan for the next decade, offered nothing to alleviate the energy crisis, and without a firm restriction on oil consumption the immediate outlook for the U.S. dollar, the spirit level of the world's economy, was grim.

Then reports of renewed unrest in Iran and its associated threats to world oil supplies shook the dollar even more. And when the dollar wobbles, the value of gold strengthens; when the dollar goes down, the value of gold goes up. Buyers were amazed this time, though, as gold rose daily and rapidly. Its ascension moved it from the financial pages of the morning newspapers to the front pages. Gradually families who knew nothing about the world of high finance realised that the contents of the trinket box in the bottom drawer in the bedroom were worth their earlier weight in gold three and four times over. Bangles and brooches given to the older ladies of Australia by beaus long-departed doubled in value, trebled, and then it was goodbye to sentimentality — down to the local gold-buyer's with them and a splendid retirement on the proceeds.

The U.S. Treasury tried to 'demonitise' gold with regular auctions of the metal, and were surprised to discover that Middle East oil sources were the big buyers, a sign that they did not relish holding their surpluses in dollars. As if the events in Asia were not shaking the dollar enough, Mr. Leonid Breznev of the Soviet Union decided to send his army into Afghanistan. The invasion was seen by dealers as the most serious development in the inflammatory Middle East situation, posing the threat of a new round of Soviet expansionism and a corresponding loss of authority, prestige and oil by the United States. Up went the value of gold again.

In Zurich, one dealer confessed that the gold market had become so big 'that it's impossible to kill it'. Buyers became convinced that the metal would play an increasing and more important role in world monetary transactions. The certainty of oil price rises, they felt, would open up more speculation into gold. There would be short-term fluctuations, but the buyers were in agreement that the rise of gold could only continue. The Anglo-American Corporation, one of the biggest gold-producers — Australia ranks fifth in the world as a gold-producer, extracting 20 tonnes a year — said it believed new gold supplies were insufficient to meet the rising demand.

Statements like these influenced individuals to decide to go out and look for their own gold. It was a simple equation: the ground held gold and gold meant money. For the outdoor-loving Australian, looking for gold was a particularly attractive occupation. The appealing aspect is that it actually works — gold is being collected from the rivers, and nuggets are being dug from the ground.

The opportunities for the layman to find gold in Australia have never been so good. He climbs into the family saloon and is at the local goldfields in an hour or so. Electronic scanning equipment enables him to sweep across ground that the old diggers had to work with pick and shovel. Metal-detectors have been developed to understand that all that glitters is not gold and can be tuned to show no response to nails, bottle-tops and silver paper.

Who in Australia needs to look for the end of a rainbow? When you hear the 'beep', start digging . . .

# 2 What Is Gold?

Chemists know gold as AU, after the Latin *aurum*. Its atomic number is 79 and its atomic weight is 196.967. Men have fought and died for it and women have cherished it. If it were to disappear from the face of the earth, the world's already wobbly monetary system would crash. Astronauts are protected from radiation by it and telephone and radio systems depend on it. It is rare and beautiful.

When you have had a lump of gold sitting heavily in the palm of your hand and you have admired its colour and texture in a way that man has done since prehistoric times, you will never mistake it for anything else. Other metals and minerals have sometimes been mistaken for *it* but when they are compared side by side, the glittering contenders fade into insignificance. The colour of gold and its consistency are unique. It is made even more important because it is not easy to find. If we were able to gather together all the gold that has ever been extracted in the world; if we were able to take all the ingots from Fort Knox, all the gold resources of the world's governments, all the watches and all the jewellery that have been shaped from it, all the gold dental fillings and all the ornate golden thrones from the tombs of ancient civilisations; if we put together in one block the total amount of gold that man has found after 6000 years of seeking, it would fit into the Prime Minister's Lodge in Canberra with room to spare.

For these reasons, gold has become man's most sought-after metal. Sometimes it is found alloyed with silver, which pales its colour; often it has a bronze tint from contact with copper. Yet it is gold all the same and man will gratefully receive it.

He has classified it as a 'noble' metal in recognition of the fact that no matter what may colour it, no substance occurring commonly in nature can destroy it. It is so malleable that the ancient Egyptians were able to hammer it into leaves so thin that it took 367 000 layers to make a pile 2 centimetres high; and it is so ductile that it can be drawn out into a thread so fine that spinners could weave a magnificent coat with it.

The importance of such a unique metal was understood as early as the sixth century B.C. when wealthy King Croesus of the Lydian Empire brought an oblong piece of gold from his vaults and gave it the name 'money'. It became quickly recognised as a medium for obtaining food and clothing. It is the only currency that has withstood the test of time, and a striking example of its importance came with the fall of Saigon. Fleeing refugees didn't take money with them; they took gold.

In its primary form gold is buried in rocks that are hundreds of millions of years old. Geologists and scientists believe it was deposited there by fluids that also forced in quartz and sometimes calcite. Thus gold, quartz and calcite are found close together, but because of the molten turbulence that occurred during the deposition gold was welded in a huge mass in some places and as fine 'splash specks' in others. Not only is it lying in the rocks around us and in the ground under our feet — it's even in the sea around the Australian coast. The total in the oceans of the world would be billions of tonnes. (The Lodge would have to be extended.) Gold is difficult enough to find on the Australian continent, but is even harder to trace in the sea: its concentration is less than 6 parts per 1 000 000 000 000 parts of sea water.

What hope then, has the weekend prospector of finding even the remotest speck?

Man has come to learn that gold is where he finds it; but some calculating doesn't go amiss. With a basic knowledge of how gold is formed the amateur prospector will recognise the most likely spots where it can be found.

Broadly speaking, gold is found in two forms: primary deposits and alluvial deposits. Primary deposits are still in the ground, and alluvial deposits are lumps and specks that have broken away from the main mass.

Primary deposits are known as epigenetic deposits because they were created after the 'host' rock had cooled following its formation. Mineral-bearing fluids — liquids or gases — were forced into cracks and weak areas of the main rock and then solidified. Because gold and quartz were forced into the fractures at the same time, it is a fairly sound indication that where quartz outcrops are found, gold is found too. However, gold cannot be separated from quartz without the combined mass being crushed. The gold is then extracted chemically.

If that were the only method available of extracting gold there would be no use for metal-detectors, pans and dredges. But nature has come to the aid of the prospector and showered the earth with alluvial deposits. Weather causes gold-bearing rocks to break away from the main face, and the gold is liberated from the stone by the abrasive action of water. That is why the only feasible prospecting open to the weekend goldseeker is alluvial but it has some rich rewards that have at times been the envy of companies committed to the expensive processes of mining.

Geologists have broken down the various types of alluvial deposits into groups and it is important that the amateur prospector should know them. First there are *stream deposits* in which gold, freed from the loose rock that contained it, finds its way into a creek or river. It is rounded and flattened by the pounding of water, rocks and gravel, and finally ends up being trapped by boulders or sinks deep into the sand to rest on the bedrock.

Then there are *eluvial deposits,* not to be confused with the general alluvial group. These are angular pieces of quartz containing gold that are found close to the main rock outcrop, usually on the top part of a hill. The quartz is also mixed with sand, rock and clay and because it has not moved far it has not been rounded or smoothed.

*Elevated deposits* are important to the metal-detector operator. These are pieces or lumps of gold that have been left high and dry after being 'deserted' by water. This may have happened after a wide stream changed the speed of its current and cut deep into its bed. Or a stream that had dried up may have started to flow again, also cutting further into the channel it earlier flowed through. Either process results in the water surface being lowered and high terraces being left at the side of the old bed.

*Hillside wash deposits* are gold pieces that are washed down a hill with sand, clay and gravel. They pile up at the foot of the hill and sometimes further up. Because it has travelled further than eluvial deposits, the gold is slightly rounded.

At least 30 metres below the earth's surface are *deep leads,* which are narrow channels containing alluvial gold. Much further up, and of importance to the pan-operator, are *leads.*

The geological terms now begin to sound technical although their meanings are simple enough. An *alluvium-covered deep lead* refers to gold-bearing sands, gravel and clay that lie between an old river bed carved many millions of years ago and the present bed. And a *sub-basaltic deep lead* is the description given to gold-bearing sands and gravel that have been covered by a flow of lava in a river valley.

Finally, there is the *elevated sub-basaltic deep lead.* Movement of the earth's surface resulted in a sub-basaltic deep lead being raised up high to the mountain tops with the hardened lava — a dark fine-grained rock — still covering it.

If you have never seen gold in nugget form or in a piece of quartz as a vein, a visit to the nearest large museum can be worthwhile. Even historical museums usually have samples and once you have seen gold you will always be able to recognise it again. You certainly won't be so mistaken that you will throw it away thinking it may be some other metal or mineral; although you may hold on to other minerals because you think they may be gold. Its colour is often described as a wax-like yellow. In natural form it is paler than jewellery shaped from it, although the colour can vary depending on where a sample is found. It is very unlikely that gold deposits recovered from one stream or surface will be exactly the same colour as those from another area. Colours vary according to the impurities in the metal. Pale gold may have been influenced by silver; darker or reddish gold by copper. The purer the gold, the higher its degree of fineness based on a scale from 0 to 1000. Much of the gold recovered in Australian rivers and at the diggings has been between .750 fine and .900 fine — 75 per cent to 90 per cent pure. Just to complicate matters, the jeweller uses the carat system of valuing gold — 24 carat being the purest. The rings and chains sold off in the early part of 1980 when the price of gold soared would have been up to 18-carat gold, the remainder of the metal consisting of platinum or copper.

It is also important for the new prospector to understand the weighing system of gold. Present-day gold-buyers are far more honest than those described in Chapter 4 and it is not suggested the 'new chum' should learn the weighing system in order to avoid being cheated. The novice should know the troy system so that when he talks to an old-timer he will understand the references to grains and pennyweights. There are 24 grains in 1 pennyweight (1 dwt). Twenty pennyweights equal 1 ounce (31.1 grams) and 240 pennyweights equal 12 ounces (373 grams). Unlike the avoirdupois method, 12 ounces in troy weight equal 1 pound.

As for the identification of natural, or native, gold, there are some simple tests. Having washed a sample, you believe you may have a piece of gold. The quickest and simplest method is to put it to the *light* test. Under a constant light source — and standing outside in the sunlight is adequate — turn the specimen slowly. If it is gold its lustre and colour will remain consistent, with no reflective spots. If you are still unsure, inspect the piece through a magnifying glass. Now you will be able to see a large number of indentations in the dull yellow surface. Gold will only have a shiny surface when it has been struck heavily, or perhaps has been hit by a spade during excavation.

If the sample is large enough, it can also be tested by *scratching*. Gold is soft enough to be easily scratched with a knife, and the resulting groove is smooth and reveals a colour the same as that on the outside. Don't carry out this test if you will be defacing a good specimen.

Another simple test is the *hammer* test. Place the sample on a flat metal surface and strike it with a small hammer. Gold will flatten out and will not splinter or crumble. It will also retain its waxy, dull colour. Whatever you do, though, *don't* use the hammer if you have a superb specimen. It's no use telling an interested buyer that the flat yellow pancake in your hand used to be a perfect miniature of the Sydney Opera House as shaped by nature.

The *weight* test can also be applied. If you are examining a nugget of gold you will notice its unusually heavy weight, particularly if you drop it into your hand.

Finally, there is the *acid* test. A drop of nitric acid on the sample will not cause it to change colour if it is gold.

Minerals that are sometimes mistaken for gold will fail one of the above tests. Iron pyrite (iron sulphide) is one of the minerals known as 'fool's gold' or 'new chum's gold'. In fact, examined beside a piece of gold it doesn't look anything like the real thing. However, in the 1850s several inexperienced diggers raced to the bank with their 'fortunes', only to have their dreams shattered when they were told they had dug up a large lump of worthless mineral. They were not familiar with the basic test of colour. Iron pyrite is a brassy yellow substance with a bright metallic lustre. It weighs less and is harder than gold — consequently when struck with a hammer it will crumble. It is very difficult to scratch and the powder is very dark green to black. Also, when heated to red hot and then allowed to cool, the iron pyrite sample will turn red.

Another new chum's gold is chalcopyrite (copper pyrites or copper sulphide). It also has a brassy yellow colour but it is darker than iron pyrite. It is not as hard as iron pyrite and can be scratched. However, the edge of the scratch breaks off into small pieces, whereas a scratch in gold is smooth. Both chalcopyrite and iron pyrites are found where gold can be traced so it is necessary to know the difference.

There is a third fool's gold — bismutite (carbonate of bismuth) — which is mistaken for gold because it is yellow and the particles tend to cling together much like gold in the panning dish. They are very dull and do not have the waxy appearance of gold.

The weekend prospector who wishes to hunt for clues to gold by studying the surrounding rocks needs to know the minerals that gold is associated with. The most common is, of course, quartz, which comes in a variety of colours. However, that which is often found with gold is white quartz which has been rust-stained. This has been brought about by pyrites within the stone. Rusty quartz, officially known as gossan, is a strong clue that gold is nearby. So is calcite, which was forced into the host rock with gold and quartz, but this is hard to identify simply because it can be any colour. It is more commonly white and sometimes forms a coating on gold. Slate is another indication that gold deposits may be found in a region and so is granite, identified by its grey colour and dark grains.

# 3 How They Found 'em, Big & Small

**Peter Culverson,** 42, grazier, Orange, New South Wales, and the late Don Peters, 50
**Discovery:** 6718 gram (216 ounce) Father's Day Nugget
**Method:** Metal-detector

'Don and I never went out prospecting without speaking to each other first and planning a route. I would ring him and we would discuss which particular spot of an area we would go over. We found the nugget south of Hill End but I am not prepared to reveal the exact location because I am convinced there is a lot more there. We put eighty to ninety hours work in on the area before we found the nugget. You don't just go out there and pick them up. For a start, you've got to know how to drive your detector. There are a lot of people going out and they don't know how to drive them. You've got to work at it. I've been interested in prospecting for some time and have done a bit of sluice boxing. I'd had the VLF–TR detector about three months and had put in a lot of hours getting to know it. Whenever I went out I knew where I was going and I knew what I was looking for. Don, too, had a pretty good idea of what he was doing. He knew the spot where we found the nugget pretty well.

'We didn't go into the area blindly — I reckon I know every rock and stone there. It was virgin country where we were looking. By that I mean it hadn't really been turned over, although people had been there before us with detectors. But as I said, you've got to know how to drive them. When I got the signal my son Paul dug in with the pick and what we saw first was a bit of dirty rock. It was jammed in tight against a root. I reckon that's what trapped it in the first place because Paul had to cut the root to get it out. We kept the nugget a week before we announced it. We wanted the area to ourselves and we really gave it a good going-over. Since then I've worked out a system with which I reckon I can pinpoint a vein of gold. I've got to test it out yet, but I'm confident it will work.

'You've just got to keep trying hard when you're looking for this stuff, but a lot of it is knowing how to drive the machine. I'm very sad to say that Don has passed away since we found the nugget, but I'm glad we were together when we discovered it.'

**Bill Hay,** 74, of Kalgoorlie, Western Australia, and son Bob, 50, purchasing officer for a nickel smelting company
**Discovery:** 3732 gram (120 ounce) Dad'n'Bob Nugget
**Method:** Metal-detector

*Bob Hay:* 'We both have a fair idea of where gold is lying around the Kalgoorlie district. You'd expect that, I suppose, after my Dad has been prospecting for forty-five years. He knows a great deal about the old methods of panning off and hard rock mining. I don't think you can buy that kind of experience or try to gain it in a year or two. You've got to get it by putting in a lot of time. Dad has owned a VLF–TR detector for a couple of years and I have had mine for about one year. I suppose you could say we are both pretty experienced at using them now. We have two different makes of detector, but both seem to work well.

'I had found a piece of specimen stone in a spot 50 kilometres north of Kalgoorlie about three or four months before. When I say specimen stone I mean a mixture of stone and gold. That was sufficient cause for me to go and peg the area. I had a chat about it with Dad and we felt it would be worthwhile to try to find what had shed this piece. What we did was make up a home-made scraping device which would be dragged along behind the car. The idea of this was to clear a lot of debris off the surface a bit at a time. We started using this device and although it didn't look very professional it worked well. We would clear a bit of land and then go over it with the two detectors.

'Well, we spent the whole day doing that without any luck and we were going to have one more go and then go home. We started with the detectors again and Dad got a signal. It was a very good signal which you would expect from a dense piece of metal. We found the nugget about 30 centimetres below the surface. The shape has a resemblance to a map of Australia.

'You have to have a lot of patience when you are searching. We were pretty tired towards the end of the afternoon after all that work with the car and the scraper but we still kept on going and you can imagine how pleased we are that we did.'

**John Aggiss,** 56, butcher, of Booradoon, Western Australia, and Frances Aggiss
**Discovery:** 4510 gram (145 ounce) Golden Beauty Nugget
**Method:** Metal-detector

*John Aggiss:* 'We were very pleased with the discovery because it's the biggest in Kalgoorlie since the 1135 ounce Golden Eagle was found in 1931. I have owned a metal-detector for four years and the first one was an excellent thing except that I found it a bit heavy for prolonged use. I decided to change it for a lighter model and I bought one for myself and one for my wife, Frances. They are both VLF–TRs from the same company, but different models. I don't want to reveal the exact location where we found the nugget because we hope to go back to the area and dig up another one! All I'll say is that it was out of Kalgoorlie.

'What happened was this: we were going to try a new spot after I had been trying one particular area without success for some time. We got to the new spot and I found that someone had pegged it out so we decided to go back to the first place. It wasn't very far at all from where we parked the car. Anyway, as we were walking back my wife said "I've got a 'beep'." So I went back to help her dig around the spot. I scraped around a bit and about 12 to 18 inches from where she had received the signal there was a little piece of an object sticking out.

'You can imagine our excitement. We could see that it was gold even though it had a reddish sort of tinge to it. I put my fingers on it to pick it up and I couldn't lift it. It was so heavily embedded that I had to lever it out. Then I could pick it up and as soon as I did I knew there was no doubt about what we had found. I tell you something, I nearly fainted. We took it over to the car and laid it out on an old blanket and then we got some water and poured it over the lump and there it was, a real big piece of gold.

'I don't want to release the name of the buyer and we have decided not to disclose the amount we got for it, except to say it was a very good price. The signal we received was very strong and don't forget we were getting it from more than a foot away. I think we would have picked it up if we had been a couple of feet from the nugget, even though it was 2 inches down. We don't regret selling it, but we've decided that when we find the next big one we'll keep that. I've no doubt that a metal-detector is the answer for the modern prospector. There is no way we could have known that our nugget was there without using a detector. The amazing thing is that we weren't seriously looking at the time, just making our way back from another area.'

**Tony Ingham,** 42, computer engineer, Mulgrave, Victoria.
**Discovery:** 187 gram (6 ounce) nugget
**Method:** Metal-detector

'I have been a part-time prospector for about five years. I hadn't been terribly successful before buying my metal-detector, a

VLF–TR. But I have put in a lot of research on the best areas to work. I enjoyed meeting people in the gold-hunting business. I read an article in a magazine one day about the arrival of metal-detectors in Australia and I decided to look at one. When it was demonstrated to me I was very impressed. I suppose I did wonder whether it would really do great things for me but I went ahead.

'I went out at weekends to areas in Gippsland because my research had shown me there were nuggety gullies there. I found a small piece of gold which must have been thrown up by someone who had been working in the creek and had missed it. I was really chuffed by that find. When my Christmas holidays came up I went to Dunolly for ten days. I only work the mullock heaps. Three of us went out and worked around the diggings, keeping a reasonable distance apart because if you are too close to another detector you get interference. We worked for half a day and didn't get anything.

'The next day I got to the diggings at 10.30 a.m. and by 11.15 I was in Dunolly having my 6 ounce nugget weighed! I had been working the mullocks around the old holes and I got a signal. I switched on to TR and got a signal which was not terribly strong. Anyway, I cleared away some earth and then put the detector over the hole to make sure I had got the object out. Then I put a little bit of soil at a time on to the detector head and then put the soil back in the hole. This was to ensure that each pile of soil did not contain the object and I was also filling the hole as I tested.

'I came to a largish rock and I thought it may have been a lump of quartz with a bit of gold in it. When I picked it up I noticed a sort of coppery colour, rather than the yellow of gold. I put a bit of spittle on it and saw the colour of gold come through. Then I was conscious of the weight. I thought "This has got to be the real thing." Strangely, I remember looking around to see if anyone else was watching me.

'I took the piece into Dunolly and had it weighed right away. At the caravan park where I was staying there was a fellow with a detector who was getting a little disillusioned because he wasn't finding anything and he was wondering if he was doing something wrong. I threw some gold down on the ground and asked him to find it, and he did. I had to keep assuring him that he was doing the right things and that he had to be patient.

**Merv Webb,** 62, pensioner, Wedderburn, Victoria
**Discovery:** 280 gram (9 ounce) nugget
**Method:** Metal-detector

'I had one particular make of detector which has a very good reputation but either it didn't like me or I wasn't using it properly. I had had it for twelve months and it wasn't doing me any good. I decided to sell it and buy another, and I heard that a fellow in the caravan park at Wedderburn was looking around for a detector so I sold him mine. You can imagine how I felt when he went out with it and within a week found a 5 ounce nugget. Anyway I bought a different make, although it was still a VLF–TR, and practised with it for a while and then went out to test it but I didn't find anything right away.

'Then I read in the paper how a girl had stubbed her toe on a 6 ounce nugget that had just been lying on the road. Well, I got to thinking about that. The road, you see, was a gravel road and what I was thinking was "Where did that gravel come from?" I had a pretty fair idea because I used to work for the shire waterworks and I knew quite a bit about the roads and the stuff they were putting on them. I reckoned the gravel was coming from Queens Gully so I went right down there that night and put the detector around but I didn't find anything.

'Anyway, my wife had a chat with the man next door who happens to be a gravel-operator and he told her that I was right in thinking that the roads were getting their gravel from Queens Gully. So I had another think about it and worked out what roads would have been treated with the gravel from Queens Gully. I telephoned a mate of mine in Echuca to tell him that I had an idea it might be worth putting detectors over the roads and he came down the next day. We were going over one road very thoroughly and that's when I got a signal and it was a very strong signal, too. I dug down about 3 or 4 inches and found the nugget. It had been pushed right into the ground by the grader and then the roller. But the detector picked it up without any trouble.

'We carried on working the area and my mate also turned up a nugget of about 4 ounces. Anyway, when I got home later I told my nephew that there was a patch of the road we had missed. He went out and he picked up a 2 ounce nugget. It makes you wonder just how much gold is lying around the gravel roads of Wedderburn, doesn't it? In fact, way back in 1928 my father found a 16 pound nugget lying on the road near Moliagul.

'I've never had a great deal of luck fossicking until I bought my metal-detector. I didn't think much of them when I had my first one because I just couldn't pick anything up with it. Perhaps I hadn't got the hang of it, I don't know. I haven't sold the nugget and I don't want to reveal where it is. I'll hang on to it. I don't sell anything, except that first metal-detector.'

# 4 'I'd Dream'd I'd Struck the Richest Lead That Ever Had Been Found . . .'

The word came down from over the Blue Mountains at the end of May 1851: 'There's a rush on — there's gold to be found everywhere.'

News of the discovery swept through Melbourne and Geelong and men's hearts began to pound. Within two weeks more than 200 workers left their jobs and wives and set out on the month's journey to Bathurst. As the wintry weather gave way, men who had never handled a pick or a spade left in their hundreds on the journey to the north-east. Small businesses with a handful of helpers were forced to close as the staff packed up their bags and headed out of town. Gold fever had set in and officials of H.M. Government and business-owners became gravely concerned about the turmoil that would inevitably follow.

'I apprehend great danger', the Honourable Thomas Icely told the Executive Council of New South Wales, 'to the residents in their own and the neighbouring districts from the great number of people who may be expected to pass through them from Port Phillip, South Australia, and the southern districts of this part of the Colony . . . From the apprehension which I entertain of the danger to which my property will be exposed I have removed my plate and other valuables, and do not intend to reside in my house. I fear that my own and all other stores along the line of road will be pillaged . . . '

In Melbourne fears of a collapse of the commercial system were so high that a group of businessmen met and agreed to offer a reward of £200 to anyone finding gold within 321 kilometres of the city. The plan worked, for it stopped men leaving the area and within a few weeks Victoria was having its own gold-rush two months before its official separation from New South Wales.

What had started Australia's first gold-rushes was the announcement of the discovery of the goldfield at Ophir, New South Wales, on 17 April 1851. Prospector Edward Hammond Hargraves, who was credited with the discovery, recalled how he travelled to a tributary of the Macquarie River and found himself in country where 'I felt myself surrounded by gold . . . '

The discovery which was soon to set the course of Australian history was later outlined by Hargraves himself:

My guide went for water to drink, and, after making a hasty repast, I told him that we were now in the gold fields, and that the gold was under his feet as he went to fetch the water for our dinner. He stared with incredulous amazement, and, on my telling him that I would now find some gold, watched my movements with the most intense interest. My own excitement, probably, was far more intense than his. I took the pick and scratched the gravel off a schistose dyke, which ran across the creek at right angles with its side; and, with the trowel, I dug a panful of earth, which I washed in the water-hole. The first trial produced a little piece of gold. 'Here it is!' I exclaimed; and I then washed five panfuls in succession, obtaining gold from all but one . . . 'This', I exclaimed to my guide, 'is a memorable day in the history of New South Wales. I shall be a baronet, you will be knighted, and my old horse will be stuffed, put into a glass-case, and sent to the British Museum!'

There had in fact been earlier discoveries, but they were hushed up. Surveyor James McBrien noted in his field book in 1823 the presence of 'numerous particles of gold' in the Fish River area, 24 kilometres from Bathurst, in February 1823. And in 1830 a convict on an outside working party found gold in the same area, but the unfortunate fellow was flogged because it was considered he had stolen jewellery and melted it down into a nugget-shaped lump. Nine years later the explorer Sir Paul de Strzelecki was requested by Governor George Gipps to keep quiet about his finds at Hartley, near Lithgow, and a later one at Bathurst, or there would be convict revolts and public unrest. When the Reverend W.B. Clarke found gold at Hartley, Gipps urged: 'Keep quiet, or we will have our throats cut.' There may have been earlier discoveries by shepherds but historians believe they did not appreciate the values of the areas in which their finds were made. The Aborigines gave the first settlers no leads because they had not considered gold attractive enough to mould into ornaments.

With a prize of £200 at stake for the discovery of a goldfield near Melbourne, false alarm followed false alarm. Hopeful prospectors trudged into town with sacks full of rocks containing the tiniest traces of gold and dumped them before the newly-appointed Gold Discovery Committee. Many of the hopefuls had brought in sackfuls of fool's gold. Finally, on 29 June 1851, James Esmond, who had previously returned with Hargraves from California, struck gold at Clunes, north of Ballarat, and shortly afterwards it was

**Off to the goldfields with all the necessary equipment — bed rolls, sieve, the cradle the sieve fits in to, spade, panning dish and kettle.**

reported that L.J. Mitchel and his party had made a find at Anderson's Creek, near Warrandyte.

There was an air of euphoria at the official celebrations on 15 July that year when Charles Joseph La Trobe, who had superintended the Port Phillip district since 1839, took office as the first Governor. The following day the Gold Discovery Committee, amid the political jubilation, announced that the two finds were genuine. It was, said the financiers, all that Victoria could have wished for — there were prospects for an economically healthy state and there would be ample jobs. But for a while many men and their families were to experience terrible hardships and, far from flourishing, many businesses crashed. Men started moving out of Melbourne and Geelong to Clunes and Warrandyte, but it was not until September, when the Cavanagh brothers arrived in Geelong with about 22 kilograms (60 pounds) of gold they had unearthed in two days at Ballarat, that the rush really started towards the north and west of Melbourne.

By December 1851, fabulously rich goldfields had opened throughout central Victoria and the cities lost their men. Flocks of sheep and herds of cattle were abandoned by farmers, doctors left their patients, and ministers of religion walked out on the scatterings of congregations they were left with. In most towns there was not an able-bodied man to be found.

The same occurred in New South Wales. 'I counted nearly sixty drays and carts, heavily laden, proceeding westward with tents, rockers, flour, tea, sugar, mining tools, etc.', wrote one Sydney observer. 'Some looked eager and impatient, some half ashamed of their errand, others sad and thoughtful, all resolved.'

From Melbourne, Governor La Trobe, in a despatch to the Home Office in London, wrote:

Cottages are deserted, houses to let, business is at a standstill, and even schools are closed. In some of the suburbs not a man is left and the women are known, for self protection . . . to group together to keep house. The ships in the harbour are in a great measure deserted and masters of vessels, like farmers, have made up parties with their men to go shares at the diggings. Both here and at Geelong all buildings and contract works, public and private, are at a standstill.

When British ships arrived the sailors walked straight off and headed for the goldfields. The owner of one vessel, in a letter to the editor of the London *Times,* lamented that every ship which arrived in Australia would find itself in the same position as his: 'namely, at anchor, with her master, and perhaps mate on board; otherwise deserted, and without hope of removal.'

Food became scarce in the towns and prices rose to many times the normal level, causing a general inflation. Flour was two shillings a pound, butter six shillings a pound, bread five shillings a four pound loaf, sugar one shilling a pound, eggs a shilling each and beer sold at five shillings a bottle.

At the diggings, where men confined their prospecting to shallow holes near stream beds and on hillsides where quartz was found on the surface, life was hard. Tempers flared over ownership of claims, fights broke out, men died from disease and attack and others prospered. It was a cruel environment, likened by many miners to a lottery where fortunes were made by nothing more than luck. But still they came along the dusty tracks from Melbourne, leaving behind them signs on their shop-fronts and homes: 'Gone to the Diggings'. They crossed mountains and rivers, some on horses and carts, others pushing barrows and all of them, man and horse, were coated in the grey-brown dust that blew into their faces on hot dry days and covered them with clay when it rained. A Melbourne journalist who rode with them noted a knot of Irishmen 'making the treeless plains vocal with their brogue and humour' and a set of 'queer fellows', evidently ex-convicts from Tasmania, who were going not so much to look for gold as to prey on those who found it.

The newcomer to the diggings was greeted by broad areas and

valleys of white tents and the ground was full of holes where men had dug frantically for the precious metal. The 'new chum' had to decide whether he should rush to stake his claim or pitch his tent, for he was allowed only a small piece of ground. Established 'residents' erected bark and log huts in which they sat on wood blocks and ate off tables made from paling. Some slept on hessian slung between two poles but most had to be content with sleeping on the ground on straw or leaves. Inevitably, lodging houses, no more than primitive huts, sprang up, where diggers were asked up to three dollars a week for a single room barely half a square metre.

'What there is to indicate gold here more than in a thousand other places that we have walked over with unconscious feet we cannot see', remarked one digger in his recollections.

Almost to a man, the diggers wore corduroy trousers, stout leather boots, a leather belt with a digging knife and a selection of hats. As there were few women at the early diggings the men didn't bother to wash or shave in the mornings. It was out of bed, to breakfast, and back down the hole.

They lived and worked with their own nationals and often whole gullies were taken over by Devonians, Cornishmen, Scots, Irish, Germans or Americans. They would group their tents together, sometimes under a national flag if anybody had one, and they would live and work as a co-operative. They all had one common driving force: the new life that a big strike could bring them. They had all privately decided it was better to share the takings of a big find with ten other men than to work alone with the chance that the man in the next hole would strike it lucky. Throughout the day it was the same routine in sun or rain — digging, sifting, searching, washing the soil. Because of the cramped conditions and the necessity to move on from hole to hole, diggers often had to walk some distance to their claims, leaving one of the group behind to cook the evening meal and guard the possessions. They often worked in parties of three to six, a large enough number to fend off any attackers of their claim — and attack was a constant worry for the 'hatter', the rare prospector who worked alone.

At first, the size of claims varied from half to one square metre for each man and in April 1853 the allowance was legally prescribed at a square metre. The rule was to work a claim every day in order to retain possession, but a miner who wished to work elsewhere or spend the day in town would simply leave his tools on his small site and hope nobody stole them. The working parties decided on their best programme, with one perhaps digging, another carrying, and a third washing and cradling. But even swapping these tasks around did little to alleviate the boredom.

A shallow deposit meant opening up the claim as a pit to expose the gold-bearing stratum, picking out nuggets with a knife — if fortune had befallen the group — and hauling the remaining rich earth to the surface in buckets. Deeper deposits meant shoring up the sides of the hole and raising the bucket with a rope. Then the excavated earth was taken to the local creek and washed in a pan. For men fired up to work fast and excitedly, as many were when they first arrived, this was a slow and wasteful method. Finally they used the puddling tub system, which had worked well in the Californian fields. The 'tub' was made from almost anything watertight, such as a hollow log with one end sealed, or half a wooden cask. The dirt was mixed with water and then gradually poured off. Water was the key to it all when men wanted to find gold. When there was no water they missed big nuggets, as so many modern prospectors fossicking around the mullock heaps are discovering to their advantage.

Each night before retiring, many diggers would fire their guns into the air to warn would-be thieves that their weapons were in good working order. Woe betide any man who tried creeping around the tents. In fact most honest men did not wander far from their tents at night in case they were mistaken for a thief. Before they lay down on their blankets or bed of straw, diggers moved the trickle of gold they had found during the day from their pockets to a bottle or a leather bag which was usually buried in the floor under the spot where they slept. As an added safeguard, some men hung cans inside the tent flap as a warning that intruders were attempting to enter.

Many of the early diggers agreed that the only punishment for a sneak thief in the night was a bullet. Those who were caught received summary punishment which varied between floggings, being tarred and feathered, being toasted over a fire or being chained to a tree for several days. Sometimes the thieves were blackballed in every camp and considered as a pariah among diggers around the country, the word being spread by the goldfield newspapers. One unfortunate thief was literally branded as a sneak for life: the words 'Caught Stealing Mates' Gold' were emblazoned on his bottom with a red hot chisel. And a cheating buyer who falsified his weights was trussed, shaved, dragged through the sludge and then flogged. Despite the obvious hatred of the diggers for those who stole, there is only one known case of lynching, described by an early writer from second-hand accounts after a digger killed another while they were dividing gold:

It was resolved to hang him, in order to deter others. The culprit . . . begged for mercy and to be taken before the Commissioner in the most abject manner but the stern reply was — 'You shall have the same mercy at our hands that you showed your messmate.' A

**Australian Gold Diggings, a glimpse at a goldfields' scene portrayed by an unknown artist between 1851 and 1860. The diggers are using cradles, panning dishes and half-casks for puddling. Some have erected winches for bringing up the earth from deep holes.**

rope was obtained, and thrown over a branch of a tree, and he was soon dangling between the heavens and the earth. The body of the murdered man was then decently interred, one of the diggers reading the beautiful and sublime service of the Church of England over his remains. There were at least four thousand diggers present at the interment and the attention and decorum paid during the ceremony had a striking effect on all present. The body of the murderer was then cut down, thrown into a hole and covered up.

Manslaughter was high on the list of crimes because there were frequent fights and robbery with violence was common. Many diggers started hiding their money rather than carry it with them but they would still be set upon by gangs who by threats and tortures extorted the hiding places. Night fossicking by candlelight on other diggers' claims, particularly those known to have made a strike, was a constant problem and so were attempts to rob on-site gold-buyers. One storekeeper who saw a hand creep under his tent one night picked up an axe, chopped it off and the next day nailed it to his counter.

Back in the cities, manufacturers were jumping on the band-wagon; those, that is, who weren't off to the diggings themselves. In Sydney, the newspapers were full of advertisements offering items like 'Waterproof tents for the El Dorado', 'Single and double guns and pistols for self defence', and 'Superfine biscuits packed in tins'. Other advertisements read: 'CHOCOLATE. Every miner should provide himself with Peek and Co.'s superior Flake Chocolate.'; 'LAVER AND CO's OPHIR CORDIAL. No one who values his health or comfort should proceed to the Gold Field without a supply.'; and 'DISSOLVE FRIGUS. As the Colony is now advancing to a state of unprecedented richness, and the empire of Australia will yet rival the age called the Golden, Leopold Morgan and Co. offer their recently compounded cordial — the Elixir of Life — which will expand the benumbed veins of the gold washers, etc.'

The rate of successful finds was not high — possibly only one man in ten could claim he had anything to write home about, and even then he sometimes lied. Those that did find gold and got it away from the diggings intact sometimes lost portions of it at the hands of unscrupulous gold-buyers. One trick was to tell the digger that the weight of gold was too much for the scales, so the pile was divided. When all the piles had been weighed separately the storekeeper would omit a grain or two, and sometimes pennyweights, when he added the total. Some sellers lost half an ounce or more. Zinc pans rubbed with grease often caused the unfortunate digger to lose some of his flour-gold when it adhered to the pan after the buyer had inspected it. Other storekeepers even grew their fingernails long so that when gathering up piles of gold dust they collected a number of grains.

Those who did make a reasonable living from the diggings made sure they made up for those long tedious hours of scratching at the earth. When they hit town the whole community knew about it. Rollicking ballads were sung in the local pubs and champagne flowed as freely as the streams at the diggings. Melbourne Town eyewitness H.G. Turner recalled:

> The Bacchanalian orgies of some of these men are scarcely credible in more sober days . . . hundreds of pounds were spent daily in reckless waste of champagne at thirty to forty shillings per bottle, ordered for the consumption of strangers and the bar loafers who congregated in the tracks of the fuddled spendthrifts. Bar counters have been washed down with expensive wines as a preliminary ceremony to ordering 'free drinks for the crowd', and with an imbecile idea of ostentation the poor fool has excited the plaudits of his followers by sweeping all the glasses off the counter with his whip to make the bill worth paying.

It was indeed a time for riotous living for the fortunate few and their hangers-on.

There was wine to be drunk and women to be had. Horses were shod with gold and bottles of the best wines were used for skittles. A digger lighting a cigar with a five-pound note would be outdone by another eating a sandwich made up of bread, butter and a ten-pound note. Publicans were quick to latch on to the fools and their gold. They kept shapely barmaids, sometimes supplied by the local madame, who enticed the men to spend and drink and then get themselves 'engaged' for a notorious ceremony known as a 'digger's wedding'. It was a pantomime performance, half mock, half real, when drunken couples betrothed themselves to one another, although it was general knowledge that some of the ladies had been through this on more than one occasion. The 'lucky' digger had to show his honourable intentions by ordering the most splendid apparel for his bride, along with the most expensive jewellery. He had to foot the bill for the wedding service and the following festivities, such as an ostentatious parade around the town in open carriages. The coachmen and horses streamed with white satin ribbons and the couple, swigging champagne and glistening with jewellery, sat back and waved to the townspeople. They would retire to the 'nuptial suite' at the old Criterion (or Shamrock) Hotel, a set of rooms splendid with gilt and mirrors for which the bill was £20 a day with a minimum of three days hire.

The population of the goldfields continued to grow as thousands of migrants came to Australia by the shipload from England, Scotland, Wales and Ireland. Most were young, single men, who came to seek their fortunes but their dreams were soon to be shattered. Letters flowed back home relating tales of bitter disappointment. 'There are hundreds of diggers returning to Melbourne every day — I may say, every hour; in fact, it is awful', wrote one arrival from the steam ship *Great Britain,* which berthed in November 1852. An Ayrshire man wrote: 'There are thousands here who cannot get situations, and great numbers are returning home. The Diggings are mere lottery. There is not above one out of a hundred that do well there . . . Several of the 'Great Britain's' first-cabin passengers are wheeling water for sale through the streets of Melbourne.'

Yet another tale of disaster was dispatched from Sydney in November 1852:

The gold fever appears to be raging all over Scotland and England at a fearful rate, so much so that I fear this country will be the scene of much misery and disappointment ere long. Young men brought up in offices should not come out. There are hundreds in Sydney now who cannot get employment; and unless they take their coats off and work at something, why they must starve. The immigration has set in upon us too fast. The country was not in a position to receive so many people so suddenly; and hence the enormous prices for houses and provisions, and indeed almost every article we use. You read in the papers of the lucky ones; no

**The ant-like activity at the Forrest Creek Diggings, Little Bendigo, as sketched by S T Gill.**

mention is made of the unfortunates: but if you were here you would both hear and see plenty about them. The lucky gold diggers cannot spend their money fast enough. Of course, the first step is to get married. The bride is decked out in the most expensive manner possible. They only purchase the highest-priced articles, and spurn the idea of taking back change. They will drive out to Botany Bay, come home all drunk, and wind up with a jolly good fight. The class of men who have generally been lucky are of the lower orders, such as labourers, draymen, bullock drivers and so on; and you can easily fancy what happens to men coming into town with a few hundred pounds in their pockets, who never before had as many shillings. They get through it all in a few weeks, and go off again. Clerks and young gentlemen, generally speaking, are the most unfortunate.

Throughout the early rushes the letters flowed back, warning 'London young men' not to come to Australia if they already had employment because, as another letter writer observed 'they can't stand the discomfort and the fatigue.' Not only the new immigrants became disillusioned at the diggings. The golden dream was fading for hundreds of 'locals', yet as fast as they left other hopefuls moved in to take their place. Prospector James Bonwick, in a guide he compiled for the early diggers, noted that 'on the way you are passed by lots of returning diggers, some of whom carry down bags of treasure and a few carrying aches and pains to the hospital.'

The reasons for the disillusion were plain. Men were not finding the vast quantities of gold everyone was talking about. They had seen the fortunate few display their wealth in the most boisterous manner and, intent on gathering equal quantities of gold, had worked hard and frantically for weeks with only a few specks to show for their toils. In addition, many fell sick, suffering a variety of ailments including rheumatism, respiratory diseases, dysentery and heatstroke. There were many deaths, but the highest incidence was at the Buckland River in 1854–5 when more than a thousand people died from an outbreak of typhoid. There were also a variety of accidental drownings and deaths from shaft collapses and falling rocks. Dysentery was the main summer disease, encouraged by flies, polluted water and the absence of any sanitary arrangements. As a remedy, men mixed a concoction of opium, brandy and chalk which resulted in prolonged constipation that was often worse than the disease. When the mosquitoes and flies were particularly bad, most people wore green veils over their faces for protection.

The disillusionment was summed up on 15 December 1852, by William Howitt, who wrote with despair from the Ovens Diggings:

> The season has been frightfully unhealthy and the journey to the gold-fields has been fatal to many. Thousands have been struck down by sickness; hundreds have already returned, cursing the parties who sent them such onesided statements of the goldfields and the climate; hundreds are still lying ill from the insidious influence of this 'fine salubrious climate'. In a letter just received from Melbourne, I hear that scarcely a soul there but has been ill and all up the country is the same. Gentlemen who have been in India, China and over the whole continents of Europe and America say that this is the worst climate they know. Without any apparent cause, people are everywhere attacked with dysentery, rheumatism, cramp and influenza. All this ought to be fully and fairly stated. Onesided statements are a dishonest procedure — 'a delusion, a mockery and a snare.' The little black fly of Australia is a perfect devil. The grass-seeds in summer which pierce your legs like needles, and actually run through the sheep-skins into the flesh of the sheep and into their lungs and kill them . . . the dust winds and the violent variations of the atmosphere — often no less than 100 degrees in a day — these are nuisances which ought to be well known.
>
> A deal is said about sending out young women to marry the men in the bush. God help such women as marry the greater portion of such fellows as the common class here! Their very language is

perfectly measled with obscenity and the vilest oaths and the basest phraseology, and they drink all they can get.

What they drank, in fact, and which added to many illnesses, were the offerings of the grog-dealers. Even in the worst areas of water shortage, breweries sprang up selling dubious liquors ranging from amber-coloured water to 'sunset-rum'. This was brewed from methylated spirit, kerosene and Worcestershire sauce and flavoured with ginger and sugar. A combination of this foul drink, heatstroke, disappointment and disease was more than enough to turn some men crazy and send them off blind with insanity into the bush to be never seen again. A local cognac was said to be made from turpentine, tobacco juice, vitriol and alcohol, and it was consumed in abundance. Some diggers were deliberately stupified with grog laced with acetate of lead, laudanum and herbs. When they passed out, they were robbed. Much of the grog was sold in primitive shanties with grandiose names such as The Royal Hotel, The Prince of Wales and The Great Southern Hotel. The more flamboyant the name, the lower the class of the establishment.

However, in spite of the tales of drunkenness, murder and robbery, the majority at the diggings were, considering their wretched environment, fairly quiet, working men whose true colours emerged on Sundays. They listened to sermons preached by wandering clergymen, and they dutifully tidied up their tents and belongings. Some would put on their Sunday best and go to visit friends and there is even a record of a cricket match between Bendigo and Castlemaine in 1853.

Men came and went from the diggings in a steady flow between 1851 and 1854, but behind the monotonous procedure trouble was stirring between the diggers and the authorities. Unable to cope with the collapse of trade and commerce caused by the rushes, the politicians attempted to invoke a high fee for a miner's licence in Victoria with the aim of forcing men back to civilian jobs.

Instead, the men continued working. Those who could not afford a licence went on as usual and when the police (the 'Joes') came on inspections, those without licences either ran into the hills or were passed licences by friends who had already had theirs checked. But the persecution continued and at a meeting on 11 November 1854, at Bakery Hill, Ballarat, the miners formulated a list of demands on the Government, including disbanding of the Gold Commission and the abolition of licences. On Wednesday, 29 November, the miners went further and voted almost unanimously to burn their licences. Foolishly, the Government retaliated by ordering another licence inspection for the next day. 'It was meant for a challenge, and as such was accepted,' wrote one of the miners. That afternoon, after commissioners and the police had been greeted with abuse and stones during the inspection, the miners met again and, wrote their leader, Peter Lalor, 'with heads uncovered, and hands raised to Heaven, they solemnly swore at all hazards to defend their rights and liberties'.

On Friday and Saturday the miners worked not in the goldfields but on building a stockade of pit-slabs on about half a hectare of the Eureka lead. The Commissioner in charge at Ballarat, Robert Reed, and a Captain Pasley who was the Colonial Engineer, were determined the diggers should be punished and they recommended to Sir Charles Hotham, La Trobe's successor, 'that sedition must be put down by force' because the diggers were fully expecting to be attacked anyway.

That Saturday night the decision was made to send in the troopers and, obviously unaware of the plan, many miners wandered from the stockade to look for food and ammunition. Others drifted away feeling that their cause had been lost anyway. Only about 150 men remained and nearly all were asleep when at three o'clock in the morning a sentinel yelled that troops and police, nearly 400 of them, were approaching. There was an exchange of gunfire and then the order was given to the troopers to charge. It was all over in ten minutes. Lalor, the miners' leader, was hit by a musket ball and by two other smaller bullets in the left shoulder. Friends who got him away from the scene procured the assistance of surgeons who amputated his arm the following day. He later detailed the casualties as 14 miners killed outright, 8 died later of wounds, 12 others were wounded and some 100 were taken prisoner. Five soldiers were killed and 12 were badly wounded.

Despite the casualties it was a victory for the miners. Licence fees were reduced from three dollars a month to two dollars a year and later to 50 cents a year. The diggers' stance resulted in a huge public reaction in sympathy with them and forced the Government to realise that the old autocratic order was unworkable. In 1855 the miners made their own mining laws and settled their own disputes. The Victorian Government set up a Court of Mines in 1857 with a presiding judge to settle disputes within the industry and this idea spread later to other Australian goldfields.

For men were chasing gold in other regions, paving a wandering path around the continent. From the goldfields of New South Wales and Victoria the hunt for gold moved north over the Queensland border and up to the Palmer River. The path then meandered west across about 3000 kilometres of blazing desert in the Northern Territory to the Kimberleys. It was slow work and again the death rate was high. The years rolled by as the trail went south from the Kimberleys to Yilgarn, near Perth. There, further diggings, more finds, more tales of disappointment, and then the chase was on again towards the east to finish in Kalgoorlie in 1893, forty-two years and some 8000 kilometres after it had begun. But what a bonanza the prospectors had found. Arthur Bayley and William Ford stopped at a native well called Coolgardie in 1892 and found 'nuggets laying about like gibber stones'. Coolgardie became the first of the goldfield towns in Western Australia and the biggest, Kalgoorlie, became the most important gold-producing area in Australia.

Men who found gold in Coolgardie and Kalgoorlie felt they had earned their prize. The hardships they had endured on the golden trail could only be rewarded with gold. Prospectors like Russian Jack, Jimmy Balzano and Dan Wicks became legends of the West for their dogged determination to strike it rich. Pushing wheelbarrows or carts containing their possessions, they walked hundreds of kilometres along the unrewarding 'pennyweight' trail. Balzano pushed his barrow more than 800 kilometres around the desert for six months and Russian Jack became famous for the record load he carried on his. The normal load for a barrow would be a 22 litre drum of water, a bag of flour, baking powder, tinned meat, sugar, tea, clothes and mining equipment. Russian Jack started from the Kimberleys with his own load and on the way to Perth he picked up the loads of two other distressed miners. Then he picked up a dying man and put him on the barrow, too.

Dying men were not uncommon and one of the cruellest areas was the Northern Territory south of Darwin. Reliable horses were rare and expensive and most men walked to the goldfields, some being attacked by crocodiles and others being speared by the blacks. In the early 1870s hundreds of men marched the 160 kilometre death trail to Yam Creek and with no transport they were unable to take much of their equipment. When they reached the jungle destination there was no gold. Many died on the site or on the return trail.

While the trail was spreading around the continent, the search for gold continued in the south-east and it resulted in a discovery that has fired the imaginations of prospectors for decades. In February 1869, two penniless Cornish miners, Richard Oates and John Deason, discovered the world's largest chunk of gold: the Welcome Stranger nugget. It was the most sensational of scores of get-rich-quick tales from the goldfields and it was among the luckiest. Oates and Deason, who had sailed from England eight years before, were on the verge of acknowledging defeat. They had heard there was gold to be picked up in the streets of Australia, but after years in the diggings with no profit to show they had become utterly disillusioned. Unable to support themselves from the mean pickings, they cultivated land as a sideline, but even that let them down when a drought struck in 1869. Up at their claim Deason, working alone, found enough water to process four more loads of dirt. Three were unproductive and with a sigh he slammed his spade into the fourth pile. It struck something solid and he cursed the 'boulder' that had jarred his bones. But as he cleared away the dirt he saw that the lump had the unmistakable colour of gold.

With a roar of delight he covered the nugget, ran to tell his wife, and dutifully sent a message to Oates. Together the two carted the nugget back to Deason's shack and at first hid it in the fireplace behind a roaring fire. Later Deason invited a few friends 'for supper'. He laid the nugget on the kitchen table and covered it with a tablecloth before anyone arrived. Then, with the party assembled around the table and thinking they were about to be fed, Deason pulled away the cloth to reveal the splendid nugget. With the amazed gasps still in their ears the next morning, the two men loaded the nugget onto a wagon and trundled 16 kilometres to Dunolly, the nearest township large enough to boast a bank.

'How much is the bank paying for gold these days?' Deason off-handedly inquired.

'Four pounds an ounce,' replied the manager.

'Ounce be blowed,' snorted Deason. 'What if it's a hundredweight?'

The bank didn't have scales large enough to weigh the nugget so it was taken to a blacksmith's shop to be broken up. The pieces weighed a fraction over 2284 ounces (71 kilograms). Deason and Oates were given the then princely sum of £9534. It is there that the story peters out and the full details of what the men did with their money have never been revealed. But the story of their amazing discovery became a legend, not only around Australia, but the world.

It is a legend that has kept men searching for gold around the country, and it is a legend that has attracted a new breed of prospectors back to the very spot where the Welcome Stranger was found.

# 5 Finding Gold: How to Start

'How much was left behind?' 'Can I really find it?' Whenever the price of gold rises, and people start to dream of making a fortune by finding their own, the doubts start to creep in. They read the history books which are full of stories of frantic digging through the 1850s and they wonder how so many thousands of desperate men could have allowed a speck of gold to remain in the ground.

The fact is they did. They left more than specks. Large nuggets were overlooked. Chunks of gold were unwittingly tossed out of holes in which the diggers searched. Large areas of gold-bearing regions were left without a spade touching the soil. And modern goldseekers with just a touch of confidence, a fair amount of luck and sometimes a little bit of know-how are now going out to collect it.

We have the advantage of knowing where to go because the old prospectors literally did most of the ground work for us. All we have to do is follow their trail. You may, of course, go your own way and try completely new areas based on your geological readings. But if you want to take the easy road to a gold mine, head the way others have gone before because there is sure to be some gold left for you. There is no need to guess; no need to take pot luck; no need to wander blindly through regions that are renowned for coal or oil-shale production and have never yielded a glimmer of gold. You need only ensure you go to a region that is known to bear gold and you have a chance of striking it rich.

If it still sounds improbable that you can go out and pick up a precious metal that man has sought for centuries, we will move in on the once-sleepy little West Australian town of Cue, 630 kilometres north-east of Perth and once the centre of one of the country's most vigorous gold-mining areas, to examine the fortunes of some modern prospectors. They are succeeding because they have invested in metal-detectors, a relatively new piece of equipment used in the search for gold.

At the focal point of the town, the old Murchison Club Hotel,

**An old prospector's cottage, like this one between Dunolly and Moliagul in Victoria, is an indication that you are not far from a goldfield.**

Mrs June Stanton the friendly proprietor will tell you that scores of prospectors have been coming into the pub to report big finds. 'One man came in and dropped a 20 ounce nugget on the bar,' she recalls. On another occasion three men sat side by side in the bar with $4000 worth of nuggets they had picked up a few hours earlier.

Cue hasn't seen anything like the modern gold rush since the old days when President Herbert Hoover of the U.S.A. once managed a mine in the area. The old Gentlemen's Club which stands opposite Mrs Stanton's hotel — and where Hoover sipped gin on the upstairs' verandah — is a decaying memory of the old days but the modern prospectors are so enthused about their finds that there is little time for nostalgia.

Among those who have picked up nuggets near Cue is Bill Harris, a former barrister who left his job in Perth to seek his fortune with a metal-detector. In his first six months of prospecting he never found less than $600 of gold a week — before gold rose on the world money market. The beginning of 1980 saw hundreds of prospectors and their metal-detectors converging on an area to the north of the town, known as the Monte Carlo Bank. This is where the first strikes in the area were made and where nuggets are still being found. At first, many of the town's 350 residents lamented the rush of modern prospectors. They have accepted it and, inspired by the successes, some have bought their own metal-detectors.

Even the local policeman has become a goldseeker. 'I've seen so many people with gold they have found I'd be silly not to have a go,' says Constable Terry Hollier. Another man who has bought a metal-detector to make his work easier is John ('Laddie') Bozanovich, known locally as the 'Old Master' because of his skill at finding gold through more traditional methods. He cursed the invasion by the metal-detector brigade, then gave in and bought his own.

'The old boys didn't get it all,' says Colin Besant, a full-time fossicker who works out of Meekatharra. Patting his faithful metal-detector he adds: 'They didn't have these things.' And Shire Clerk, Geoff Foster, grins: 'Goodness knows what the old-timers would have uncovered if they had used metal-detectors a hundred years ago. From the amount of gold around here they would have needed an armoured vehicle to cart it away.'

That the gold is there for the taking is revealed by the fortunes of Trevor Eyden, who was visiting a relative in Cue when he decided to test his metal-detector near the town's caravan park. It beeped over a piece of stone which turned out to be a nugget of 171 grams (5½ ounces) and almost solid gold. There are also the fortunes of the brothers Thompson to consider. In the last few months of 1979 Murray and his brother Ian, from South Australia, found gold worth $25 000.

So gold is out there. And it is not just sitting in the desert near Cue. It could be lying in a forest in South Australia, mingling in the dirt of a Victorian hillside, sitting at the bottom of a Queensland river or resting at the top of a New South Wales mountain. How then do you find these gold-bearing regions?

Specific spots to try in each state are described later but your own inquiries could begin at the Mines Department — the name varies from state to state — or at a library. It really depends on how much work the Mines Department has done in the preparation of maps detailing the goldfields. You will certainly be able to purchase a general 'mineral' map of the state which will have gold-bearing regions marked. The undisputed top of the hit parade for nuggets is the north-west area of Victoria and prospectors in that state can benefit from an immense amount of work that the Department of Minerals and Energy has put in on their behalf. You can purchase a state map specifically detailing the goldfields and also a detailed map of any particular region you wish to try. You are also offered, for $50, a package deal of thirty-six maps of Victorian goldfields, a Miner's Right certificate — which you must have — and an edition of their *Prospector's Guide*. Although dated, the booklet's contents are still relevant.

Armed with your state map, you would now like to know which goldfield to head for to find gold. If any of the experts at the Mines Department could tell you that with any certainty he wouldn't be sitting in the Mines Department. As the prospectors of old said so often as they returned to Melbourne, Ballarat or Bathurst with blistered hands and empty pockets, gold-hunting is a lottery. Your lottery chances are estimated by experienced prospectors as anything between ten to one and fifty to one, providing you prospect in areas that have produced gold. You can find this out the slow, but more interesting, way by reading some history books at the library. Many of the books written about the gold towns simply use the name as the title, for example *Dunolly, Story of an Old Gold-Diggings* or *Old Walhalla – Portrait of a Gold Town*. Many of these types of books give details of the amounts of gold discovered and sometimes you can establish whether your intended search area produced surface alluvial gold or deep-lying reefs that had to be mined.

The alternative method of finding the best places to choose is to ask at your Mines Department for a list of the yields in the state, perhaps with details for particular areas. Some departments can provide photocopies of reports from the old Mining Registrars who detailed the finds and the activities on the goldfields.

Having arrived at the goldfield of your choice, you will find yourself in a region of slate or granite outcrops. You may also notice some hard siliceous rocks like jasper or quartzite. This, then, is auriferous, or gold-bearing, country. The first thing to do, like any smart businessman, is to investigate the opposition's moves. If there are any old-timers around — and don't laugh because you *can* still find them at the diggings — don't just give him a 'G'day' and plunge blindly off up the valley. Ask him where you should look. He won't tell you his favourite spots but he will almost certainly be good enough to point you in the right general direction.

You should also talk to the modern prospectors, who will be happy enough to chat because of a form of brotherhood that has been created on the goldfields — particularly among those with the same make of metal-detector! Undoubtedly those who haven't found gold in the area you have come to will announce it is hopeless. It has, they will try to tell you, been 'done over' by past and present goldseekers. But the next day someone close by will have found another nugget. Don't be put off by the seemingly endless holes in the ground, nor by the piles of loose material beside the creek, all of which suggests that the area has been turned over by generations of miners. These are positive clues that gold is nearby. You should only start to question your choice of area — unless you are very experienced at recognising the characteristics of gold-bearing regions — if there are no signs of prospecting.

It will help your search if you know how gold is distributed. Many of the old prospectors dug in next to the other man without considering how or why gold came to be there. But a few minutes spent absorbing the following facts may lead you to spots in the gold-bearing regions that the 'new chums' ignored.

At the start of the alluvial journey, gold-impregnated rocks are carried down the hillsides by water and gravity and as they tumble they break up, spilling out the gold. If the gold does reach the river, it will sink to the very bottom, below the sand. Sometimes it becomes trapped high on the hill in a depression. It could be just about anywhere and associated with almost any type of rock.

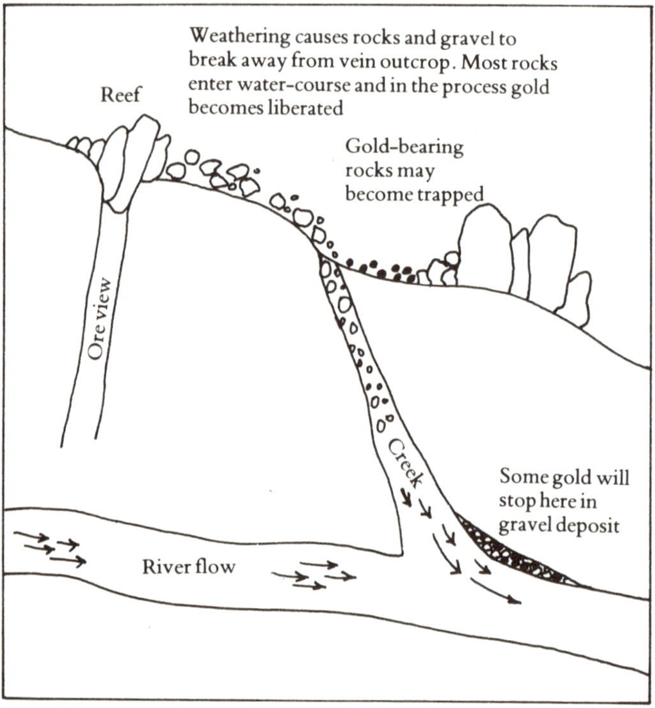


**How gold ends up in your panning dish or under the search-coil of your metal-detector. It breaks away from the main reef and tumbles down the hillside. It may become trapped behind boulders or continue along a creek and into a river.**

The initial breaking away from the host rock is caused by weathering. This can occur in three ways: by *water* changes, *temperature* changes or *living nature* changes.

Water changes are sometimes known as chemical changes but the effect is still the same. Water causes rock-bearing gold to break away from the main mass. This can be brought about by the rock taking in water through a crack, swelling and breaking away. Or it can be caused by rain or a creek simply wearing the rock down so that it is washed from the main face.

Temperature changes occur when the rock takes in the heat of the sun all day and is then subjected to cold night air. The expansion and contraction of different minerals within the rock face cause the stone to break up under the stress.

Living nature changes are caused by the roots of plants which force their way into rocks through cracks, expand, and cause parts to break off. Or such changes can be caused by lichens which produce rock-attacking acids, or animal droppings which can also have a chemical effect.

There is a fourth reason why gold-bearing rock breaks away: *erosion,* not to be confused with weathering. When erosion occurs, winds and water carry abrasive stones over the rock face and wear it down, rather like the effect of sandpaper on a piece of wood. Erosion occurs after one of the three weathering effects has taken place. In other words, when chunks break away from the main rock face and tumble down, they break up and help the wind and water to rub away the surface of the main mass.

If the main rock face has been holding gold, then it follows that the chances are high of gold being broken away by one of these four processes and carried down over the hill by water, gravity, or a combination of the two. Water does most of the work. Heavy rains will wash the rock breakages towards the valleys and gullies and into streams. These streams continue to wash the rock parts onward to the sea. Gold, however, because it is nearly twenty times heavier than water, will tend to become lodged behind the most simple of barricades whereas the lighter particles of stone will continue to be washed on. Being so heavy, gold will settle down through sand and gravel until it reaches the point of impact — in this case the hard bedrock to be found under the gravel and sand on the river bed.

Gold which does not get barricaded will continue to be dragged on by the water and, because of its malleability, will become rounded and flattened. The more rounded and flattened it is, the further you are from its source in the hills. If it is very sharp and jagged when found it could be lying a few metres from a rich primary deposit.

On its journey down the river, gold can be dumped on bends where the 'inside lane' of the current, because it is travelling more slowly, will not be able to keep the yellow metal suspended and moving. Should the gold escape the natural river barriers and be carried around bends by fast currents, it can still be held up and deposited in places where the stream widens and spreads out like a

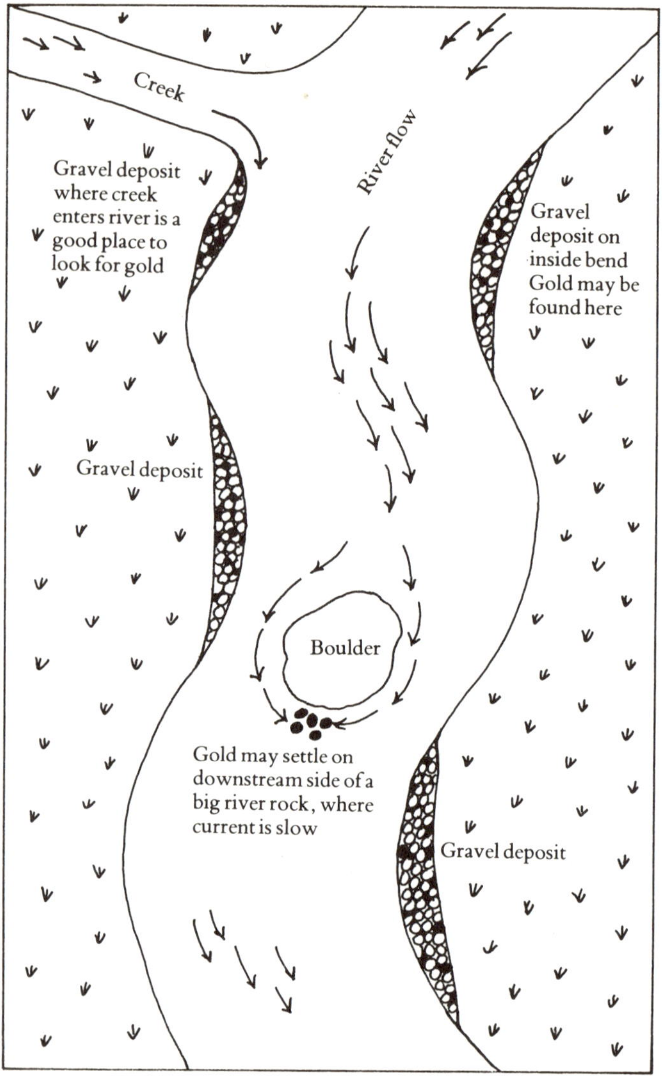


**Once it leaves the fast-flowing current of a creek, gold will settle in slower-moving water. This could be as early as the junction where the creek meets a river. Rounded river boulders which slow the river current may also hold gold on the downstream side.**

webbed foot onto flat, low-lying land. This causes a slowing down of the current which results in gold settling down into the sand. With the slow current moving above it and the sand continually shifting, gold snuggles into the sand, searching out its final resting place on the underlying bedrock. Here, it nestles into cracks in the stone and in Chapter 6 (on panning) the importance of getting down to this bedrock area will be emphasised.

Gold can be found behind any natural obstruction either in or out of water. If a river passes over a small ledge, gold may have dropped in a depression that has been formed by the water, but if the current is strong enough it will be dragged on again. Its movement is erratic, giving the best and worst prospectors an even chance: you can be absolutely brilliant and pick up its trail or you can be a novice, take pot luck, and turn up as much gold as the expert. You shouldn't go wrong, though, if you regard every nook and cranny in the river bed or in any boulders in the water as gold traps.

Areas where the current slows should be thoroughly investigated. In addition to bends, this could be where a fast-flowing 'feeder' stream meets a slow-moving river. Gold will be carried into the flow of the slow river, but will soon settle because the current is too weak to help it on its journey. The spot to look is 'just around the corner' on the downstream side of the main river.

Why then, if gold always falls downwards, can we expect to find it on a mountain top? This has nothing to do with Australia being 'upside down' as some of our Northern Hemisphere friends may jest: it's all to do with geological movement which took place untold millions of years ago. Gold lying in river valleys has been carried up to the hilly regions by a remarkable shifting of the earth's surface. Gold has either stayed up there, been scattered by the buckling of the earth's surface, or has been lost under a river of molten lava which has set into hard rock — a safe deposit box which nature may unlock in a million years or so.

Gold-prospecting, either with a metal-detector or by using more traditional methods, is like playing detective. Nature provides you with the clues: you follow them up. There may be a trail of fine gold, a *lead,* in a stream which you can trace up to the hills. You may find a *float,* which is a broken-off piece of rock containing gold and by finding one you may be able to trace another further up . . . another trail which, if you can pick it up, can put cash in your pocket. More clues can be found by *loaming* — gathering samples of earth on the side of a hill and, if gold is found, taking other samples to establish the trail of gold.

You will have gathered that the gold detective's greatest asset is observation. Once you have found gold in one spot and got the 'feel' of your surroundings, you can play out your hunches in other areas that seem familiar. If, for example, you found samples at the base of an old tree-trunk in a dried-up valley, you will be paying particular attention to the vegetation in the next valley. Take notes,

**Pan wherever the current slows down. This Victorian prospector actually found gold in the still water behind this log.**

if necessary, of the characteristics of the spot where you found gold. If you found gold on the downstream side of a small island, where the current is slower, that again would be another place to check at the next river.

You will soon learn the obvious places to look. On the other hand, you will never know the obvious places *not* to look in a gold-producing region because it could be just about anywhere.

# 6 Panning: Where to Go, How to Do It

### Wet Panning

*Equipment necessary:* One panning dish; knife; tweezers; magnifying glass; various lengths and thicknesses of stiff wire (coathanger pieces are good); shovel; trowel; wellington boots; glass bottle.

No matter what advances man has made in the detection and extraction of gold, the one implement that has remained close by his side through the centuries from the days of the ancient Egyptians is the humble gold-pan. Its shape may have altered from a bowl, to a bottle, to a dish, but the modern implement is worked on the same basic principle applied thousands of years ago. Without it, vast quantities of gold would have been left in the world's rivers. It is used to separate gold particles from river gravel, hence the use of the expression 'panning for gold'.

Some prospectors are masters of the art. Others, by clumsy use of the pan, have probably put back into the river all that they took out. It is imperative that all prospectors should learn the correct way to pan because even the metal-detector has to be put aside when a signal is received from 'somewhere' on the river bed. This chapter, based on the experiences of men who have been using the system successfully long before metal-detectors were introduced, will teach you the right way to pan and where to search for gold when you have only the dish, a knife and a spade (and perhaps a few other items from the list above) as your companions.

Two types of pans are available in Australia. The difficult one for beginners is the traditional metal type, usually steel, although some earlier ones were copper. The old prospectors will turn over in their graves at the knowledge, but the second and more efficient pan is made of *plastic* — green plastic. There's a reason for the colour, to be explained shortly.

First, the steel pan, faithful servant for millions of prospectors around the world, from bearded old-timers with floppy hats to the modern goldseeker in his T-shirt (although in Walhalla, Victoria, there is a combination of the two: an old-timer in a T-shirt). 'New chums' to the diggings often had to learn to pan the hard way after picking up their dishes from the local blacksmith. Competition was keen in those very early days and not too much information was given away. The crudely-shaped pans retained very few particles of gold in the hands of the inexperienced, although they were cleverly substituted at the end of the day to serve as cooking pots.

The most widely-used steel pan in Australian rivers is between 40 and 44 centimetres around its top edge and about 30 centimetres circumference at the bottom. It's about 8 centimetres in depth. You can buy smaller ones measuring 15 centimetres across the top or a very wide version 50 centimetres across. Some 3 or 4 centimetres from the lip of the average pan is a 'riffle', an indented strip designed to stop the gold falling from the dish during the panning process. The idea is to catch the gold in this strip as the sand and gravel is gradually swished out of the pan, but prospectors admit it takes practice.

Before you can use the steel dish, you must 'blue' it by sitting it on a glowing fire until the metal changes colour with the heat and then dropping it into water. If unsuccessful the first time, you may have to repeat the process. The purpose of this is to remove a film of oil which has been applied to prevent rust during pre-sale storage. If you fail to remove the oil, gold specks could become stuck and may be washed out with the film during panning. Blueing has to be carried out at intervals during the year to remove any oils that have come from your own hands. This is a very necessary process that would make a trip to the river pointless if it wasn't done. Anyway, the darker background of the pan helps you spot the waxy yellow colour of gold.

The plastic pan made its appearance in Australia at the end of the 1970s. Its general shape is the same as the steel version, but it has a series of inside 'ledges' from the top edge to the base, rather like terracing. Those ledges, or riffles, with their ninety degree angle, serve as excellent traps for fine pieces of gold. If you had nothing but water and gold in the pan and tried to pour the contents away over the terracing, the gold would be trapped in one of the angles.

The plastic pan, lighter than the steel version, has an additional advantage: the green colour. Modern researchers have found that green shows up gold much better than steel, even when the metal has been blued. The green, however, has to be just right. Too light, and the sun reflects off it, making the gold difficult to see. Too dark, and it takes on a black appearance on overcast days, making the tinier particles of gold almost impossible to spot. Researchers have tested this colour successfully under all conditions.

The plastic pan is, of course, rust-proof and non-corrosive, although it, too, has to be cleaned of body oils occasionally with a rubbing alcohol. One of its other advantages is the plastic consistency, which can be given a rough surfacing on the inside base of the pan to help trap dusty particles known as 'flour gold'. The plastic pan's only disadvantage is its performance if you leave it under the back window of your car on a blazing hot Australian day: you could have something akin to a slimy space alien sliding over your back seat.

The idea of panning is to wash away all the stones and gravel from the dish in order to leave gold lying at the bottom. This is all done with gravity and water, gravity playing the greater role. Gold is 19.3 times as heavy as an equal quantity of water and about 8 times heavier than the sand and gravel you pour into the pan. Because of its weight, gold will always try to get to the bottom of things. It will tumble, through weathering, from the top of a mountain. It will go to the bottom of a river bed, except in unusual circumstances which will be explained later. And if you pan correctly it will work its way to the bottom of your dish.

All you have to do is work the way nature does . . . slowly. More often than not, nature uses water to shift gold — hence old prospectors used rivers as their base. They also needed it for washing sand and clay from specimens they had collected. (On occasions they were even known to wash their shirts in it.) When you confine your search to rivers and gullies you are looking for *placer* gold (pronounced plasser) which has been carried downhill by water and gravity to the river.

To start your river-prospecting you need to pull on your rubber boots and wade out to the middle of the stream or creek with your spade because that is the best place to start searching. Gold, because of its weight, follows the deep channel in the centre of the stream. If the water floods in over the top of your boots and is too deep to work you will either have to forget this spot, build a wing dam, which is not legal, or invest in a dredge (see Chapter 11). At first you should look for big rocks or boulders that run in an approximate line from bank to bank, or tree roots or old logs — anything that is creating a damming effect. Rocks that are part of the bedrock are a 'bar'. Small pebbles and stones will have been washed up against them and in among these some gold may have been trapped.

However, you still aren't ready to start your river-prospecting yet. The most important thing is to establish the depth of the bedrock at the dam or 'bar'. The bedrock is the true bottom of the river, the really hard rock that lies under the sand and gravel. If you strike it with your spade in the first 30 centimetres or so, fine. If you're still burrowing with your head under water, find another spot where the bedrock is not so deep.

So we now assume you have found shallow bedrock and you are

standing at the bar in the middle of the river. Pick up all the pebbles and stones from the upstream side of the dam, concentrating on the buried ones, and inspect them by hand. They will have been thoroughly washed and there might be a nugget among them, although there aren't too many nuggets in rivers nowadays. Next, dig your spade deep into the gravel as far down as you can to the bedrock. If there is clay just above the bedrock, be sure to gather this up because gold often becomes embedded in it, brought about by the pressure of the sand above. When you have a spadeful, wade back with it to the dish, which you should have left on the bank.

It's too difficult a procedure for the inexperienced panner to wade out into the stream with his pan and his spade and expect to be able to dig down with one hand while holding the dish in the other. You can, of course, balance the pan on one of the mid-stream boulders, but is it really worth the risk of it tipping over?

You should aim to fill your dish by about half or three-quarters so you will probably need to shovel up more from the river bed. You must really try to scrape the bottom with the spade because gold sometimes becomes cemented to the bedrock by clay that has become almost as hard as the rock itself. Get into as many nooks and crannies as you can; scrape and dig, disturbing everything that is on the bedrock. By doing this you will disturb gold that has embedded itself. When you bring the spade and its load up move very slowly because the flowing water will try to steal the contents of the spade. You should now have enough material to pan. You

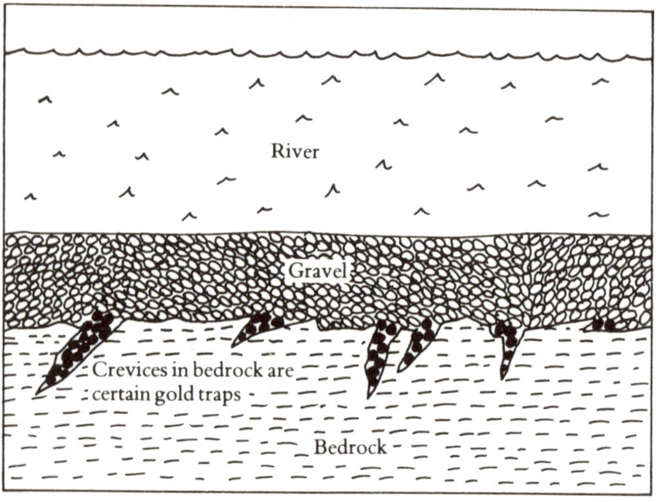


**The bedrock under the river-bed gravels is heavily creviced in slatey country. Gold will always settle into these cracks and is best reached by gouging open the slate.**

will return shortly to the boulders to do some picking with pieces of wire. Some old prospectors suggest you should pick at the cracks in the boulders first, putting into your dish all the particles you find, and then dig into the creek bed. However, the majority say that by digging first and picking the cracks second you have a chance of finding any particles of gold you might disturb with your digging and which are carried by the current into the cracks.

So you have now waded back to the edge of the stream to pan the diggings from the foot of the bar. You should look for a working area, where the water is fairly shallow but deep enough for you to be able to totally submerge the dish. If you can find a rock to sit on, so much the better. Be warned, though, not to select an area where the water runs fast because this will make your panning extremely hazardous. It could cost you a few hundred dollars in washed-away gold no matter what type of gold-pan you are using.

The first step in panning is to get the contents of the dish thoroughly wet. Only by doing so can you ensure that the gold works its way to the bottom. Immerse the pan completely under the water by about 3 centimetres and stir the contents with your finger (you'll be able to hold the dish with one hand: if it's too heavy, prop your foot or a rock under it for support). The stirring helps to free the gold from the gravel and it will then fall to the base of the dish. If you see bubbles, that means you are breaking up the clay and releasing air pockets. Wash the bigger rocks first, inspect them and discard them. Unless of course they are worth keeping . . .

Now stir again. What you are doing is making absolutely sure that particles of gold are not caught up in any clay or slime. By getting the clay saturated and then breaking it up, liquifying it in fact, you are freeing any gold that was stuck to it. Gold will move to the centre of a circular flow of liquid and, as the flow slows, will settle on the bottom of the dish.

You are now at the 'lapping' stage, when you can start to spill off the lighter, worthless, contents by allowing the water to collect it and wash it away, like the tide reclaiming sand from the beach. Experts are varied in their opinions here whether the dish should remain totally submerged or whether the edge farthest from you should be the only part to stay under water. If you use the latter method, keeping half or a third of the dish out of water, you must ensure your tilt is enough to allow water to lap back and cover the dish contents because they must, at all times, be kept in water. Whatever method you use you must keep the water swirling. So, if you decide to keep the dish totally submerged, you should rotate the contents by grasping the pan on the left side and the right side and swirling it (an anti-clockwise motion is best for right-handers). If the dish is kept partly out of the river you must rock it

Kevin Cervi, 28, from Willow Grove, Victoria, demonstrates his art at panning with a metal dish. 'You just have to take it very easy,' he says. 'Why take unnecessary risks with your gold?'

away from you and then towards you so the water laps over the gravel.

If you are swirling the dish you must tilt it occasionally to allow the lighter material to be gently washed out. If you are rocking, the light stuff will be 'lapped' out. All around the dish the water will become discoloured, which means you are doing a good job in breaking up the clay.

When the water in the pan clears, and not before, you should shake the dish to free some of the smaller stones at the bottom. As you shake, either towards you and away, or from side to side, the action will work these smaller stones towards the surface of the gravel and you can 'rake' them off with your hand. Don't worry about raking off any gold. It will be safe at the bottom of your pan. You can also start to rake off the bigger pieces of gravel. But you must continue the swirling and shaking of the pan each time you rake away a few more stones. Continue the spilling, shaking and raking until the dish is about a third full. It is now time to be careful. Allow the river to lap away at the material while you tilt the dish slightly until you start to see sand peeping through the rubble. When you do, bring the dish level and shake it to resettle the contents. Spill out more gravel and rubble until the sand shows through again, and once more regroup the contents by shaking the pan on a level keel. These instructions may seem very detailed and the process may appear slow. In fact, when you are panning, it will all seem very simple and you will find that because you *know* what you have to do — that is, keep the heavy gold at the bottom of the dish — many of these actions will come naturally.

You may wonder at this stage, with so much raking and shaking, how much gold has raced away in the current. If you have been gentle and patient you shouldn't have lost any. However, if you did not keep the contents thoroughly wet, and of a consistency similar to soup, clots of unbroken clay may have sloshed out with some of your earnings. If you use a plastic dish with its inbuilt riffles and you have stirred the contents well and spilled out gently, then there is nothing to worry about. You will lose gold only if you have operated clumsily, or if you are using a steel dish which you have not mastered. Some experts with the steel dish spill out their lighter material by immersing the pan and bringing it up quickly, allowing the action of the water to carry the material over the edge of the pan. Certainly you may practise this but use material from near the river bank. Don't take the risk of practising with gravel that could contain gold (actually gold can be found by the river bank, but that will be described on page xx).

When you have lapped and tilted away the bigger pieces of gravel and you are down to the sand, you must exercise great care. There should be about a mugful or so and, if you are lucky, somewhere in among it is gold. It will either be in grains or fine flour gold. If you had anything bigger, it has gone because you did something wrong. There is a chance that the fine gold has not been able to sink through to the bottom of the pan because of the tight consistency of the sand. This is why you must now ensure that the contents are under water and that you thoroughly rock and swirl the dish, giving the gold the chance to work its way down through the sand. If you do not keep rocking and rotating, keeping the contents suspended in water, you may lose gold because even the sharply angled terraces inside the plastic pan will not stop a ball of clay containing gold particles. In fact, if much of the material in your dish is of a clay consistency you must be very particular to break it down at the outset to a thin watery paste. This is known as *puddling*. Submerge the pan and stir the clay with a stick or your finger to allow the current to wash it away. Apart from 'freeing' gold particles, washing away the clay ensures that the water in your dish is clean to enable you to see the contents.

It is now assumed you have washed everything from your pan except sand and gold. At this stage you need have no more than a couple of mugfuls of water in the dish and you can start to pour it and the sand over the terraces in the plastic dish. If you are using a metal pan you must take extra care and use the 'lapping' method to remove the sand. You do this by tilting the dish away from you at a steep angle and, as you raise and lower the far edge, you allow water to seep in and gently remove small amounts of sand. Stop this process every two or three seconds, bring the dish level, and rock it from side to side to ensure gold is staying at the bottom. Whatever dish you are using, you should have now poured away enough sand to be able to see whether you have any gold. If you cannot see any it doesn't mean it isn't there. To make sure, give the dish, containing about a quarter of a cupful of water, a quick swirl, stopping it suddenly. This action will throw the sand ahead of the gold, leaving the yellow metal behind in the 'tail'. The swirling method was used by the old prospectors to complete each panning and, whether a plastic or a steel dish is in use, that is how you must finish.

If you see gold particles, pick them out with tweezers or a moist finger, and put them in a small bottle of water which can be sealed. If you have picked up sand with them you will be able to see it and the gold clearly. At home you should be able to pour most of the sand out and separate the last particles of gold and sand on a piece of plain white paper.

Should you have any doubts that you have picked up all the gold specks from the dish, you can carry out one more manoeuvre to satisfy yourself. With just a few spoonfuls of water in the dish, roughly equalling the quantity of sand, tilt the pan so everything is in one corner. Then shake the dish by rocking each hand up and down, slowly rotating the pan as you do so. Any flour gold remaining should trail out behind the particles of sand. If it's not clear, use your magnifying glass and pick up any specks with a moist finger. At this stage even tweezers may be too clumsy to use.

So you have now panned, and perhaps collected, from your diggings in front of the creek bar. It's now time to check out the cracks in the rocks themselves. Using a knife, a piece of coathanger wire, a garden trowel or any appropriate 'scooper', hook out from the cracks and crevices any tiny stones or gravel trapped there. You can search on the rocks above the water-line and below because if this is a very old river the gold could have been distributed at various levels as it changed the nature of its course. You may be able to prise some of the cracks apart, particularly if they are in slate, with your spade or trowel. Remove from the fissure *everything* — clay, small stones, flakings from the slate. If there are stones you can't get to above the water-line, one trick is to put clay on the end of a stick and use that to pick them out. You could use sticky tape on the end of a stick or a piece of wire, too. Some of the small stones will have been washed into the cracks when you gathered your first panload from the river bed earlier. If you found gold from that exercise it is likely that some of the gravel you lost as you brought the spade up through the water contained more gold. It's an important operation, then, to dig out all the cracks. You can now pan these, but greater initial care should be taken because you will be working with finer material with which gold particles can mix. Just remember to keep the contents in plenty of water as you rock, swirl, rock, swirl, allowing a little dirt and gravel to lap occasionally over the edge of the dish.

When you have finished with the material from the cracks in the boulders, don't charge off up to the hills to start hunting in a completely different area. When you are on to a good thing stick with it. Wade out to the *downstream* side of the bar and start digging there. It is possible that some gold may have got through this natural barrier but because it has been slowed down has become trapped behind a minor obstacle at the bottom of the river. In addition, water pouring over the top of the bar tends to form a depression on the downstream side. This, too, can be a catch for gold. If you can find quartz and ironstone in the depressions, you can also find gold.

Having panned the downstream area you are now ready to move upstream from the bar and take samples from near where you first started. What you have to remember when you dig is to get below the surface of the creek bed. Take no notice of those fellows you see scooping sand from the river bed without really digging down. It's very unlikely, though not impossible, that they will find any gold in the surface of the gravel and sand. It means more work to dig down to the solid bedrock, but you can't expect to make a fortune without putting some effort into it. Over the years the heavy gold will have worked its way down through the sand on the river bed to the granite base. It will have been assisted by swift-flowing currents which made its movement through the sand easier. If that action sounds similar to panning, that's exactly how it is. The prospector with his pan is copying nature by giving gold plenty of water, plenty of movement, to filter down to the bottom of his pan, just as the metal settles through the river-bed sand to the bedrock. That is why you should be concerned with only the bottom few centimetres when you scoop into the sand.

You have now been digging about a metre up from the bar of boulders. Perhaps you have found more gold. Perhaps you haven't found any since you started. If you have, during a few attempts, keep your phone number private because you have the kind of luck that everyone wants to talk to you about.

When you have worked the area near the boulders thoroughly and have found gold, move about a metre upstream. If you haven't found any, move up to the next bar of boulders and try the cracks there. The old prospectors wouldn't bother to stay in the same spot if their work yielded only a speck or two. They would consider that unpayable and move upriver, perhaps as much as a kilometre or so. However, what they would consider unpayable would be worth seeking, considering modern gold prices.

If you have found gold you can be sure there is more upstream. Gold in rivers runs in what is known as a 'lead', sometimes 3 centimetres wide, sometimes much less. All you have to do is follow it towards the hills. The fact that gold is in the river means that it has been weathered or eroded from a reef in the hills and has found its way into a gutter on the river bed. Like a fine thread, however, the lead sometimes gets broken and it could take an hour or two to find it again. The gaps will have been caused by fine particles of gold becoming caught in very thick clay, while other pieces were able to flow on past to continue the trail further downstream. You have to dig, pan, dig, pan. Above all, persevere. That's what prospecting is all about. In fact it is harder to persevere nowadays than it was for the old diggers. You don't *really* need the money; the wife and kids aren't sitting in a tent down in the valley wondering whether you're going to turn up enough grains to buy the next pound of flour for the daily bread. You've also got the stationwagon parked just over the hill and it's tempting to pack it all in for the day and drive down to the pub. All right, so you've only got a bicycle, but it's still more than the chums at the diggings had and still a lure from all this backbreaking work.

To be truthful, some of the early prospectors did find gold easily in the creeks. Here is a description of a prospect at one river by William Howitt:

The creek runs about fifty yards to the left of our tent, and is, perhaps, some ten or a dozen feet across; but the stream only occupies part of this space, running among thickets of the aforementioned shrubs . . . This is what may be called surface-digging, for the gold lies near the surface of the bottom of the creek; and, indeed, we find it hanging in the roots of shrubs that we pull up from the spaces that stand above the water. But the main quantity of gold in this stream is found to lie on a layer of clay about two feet below the surface. On this lies mud; and in this mud there is more or less gold. We have no need, therefore, to dig holes of from ten to eighty feet here . . . Imagine us setting to work on any brook that you know . . . It would seem absurd to expect gold in it. But there is nothing in these brooks that looks different: the mud is just as black and muddy, and the wet and dirt are just as wet and dirty. Yet out of the very roots of the grass we shake gold. We can see the particles shining as we open pieces of the grass-roots, but the chief part of the gold lies on the clay at the bottom, and that seems studded with it just like a pudding with plums.

Nature has been kind enough to leave the alluvial prospector with certain clues. Once the goldseeker knows what to look for he is halfway there. If you can detect one clever little game that nature plays, you can pick up more gold than you bargained for. You may be wading upstream with pan and shovel, following a thin trail of gold specks, chasing the thread that you have picked up at the rock bar. But beside you is a *second* lead, 3 centimetres wide, that you don't know about. There may be even three leads. Many early prospectors lost plenty of money because they weren't wise to this trick, but we can gain from their losses. The way to find out if there are two leads — and indeed to find one lead if you have no luck near the bar — is to dig across the flow of water, from bank to bank. You have to be extremely careful and be sure to remember when you start dishing the material from the trench exactly where each panful came from. It's no use piling in a dishful of gravel from the first lead and then pouring in material from another part of the trench that might contain the second lead.

**Ideal panning country. The swirling currents in this river-offshoot will throw gold into the calm spots among the rocks. The roots of grass may also trap gold. The bedrock above the water surface has plenty of nooks and crannies which should be investigated because gold may have been deposited when the river level was higher.**

No, first you must be sure you know where the first lead is and mark the spot with a stick driven into the river bed. Then you take another panful from close by and examine that. Every panful must come from one particular spot and you should also mark that spot with a second stick. When you have found the second lead, if it exists, you will know exactly where it came from. If the river is narrow enough and you have not found either a first or second lead in the trench, you may feel like starting another trench. But it is hard work and one trench in a day may be more than enough.

As you follow the lead, you may find that the gutter it has been running along veers into the river bank. This is not the end of the trail. The gutter probably continues across dry land. Once it would have been flowing with water, bringing gold along with it. The gutter may now be overgrown but if you can trace it and dig down you may be able to pick up the trail of gold again. Having noticed the characteristics of the overgrown gutter — it may be slightly shallower than the surrounding ground or it may contain a different type of topsoil — follow the clues on up the hillside. Dig in and pan occasionally to check that the lead is still running. This is where you will find dry panning of assistance and this will be explained in detail in this chapter.

Perhaps we should pause here while you ponder over the golden trail being discussed. No doubt you are asking: 'How can I, as an amateur prospector, possibly hope to pick up a trail of gold from a river dam, follow it upstream and then go chasing it across the

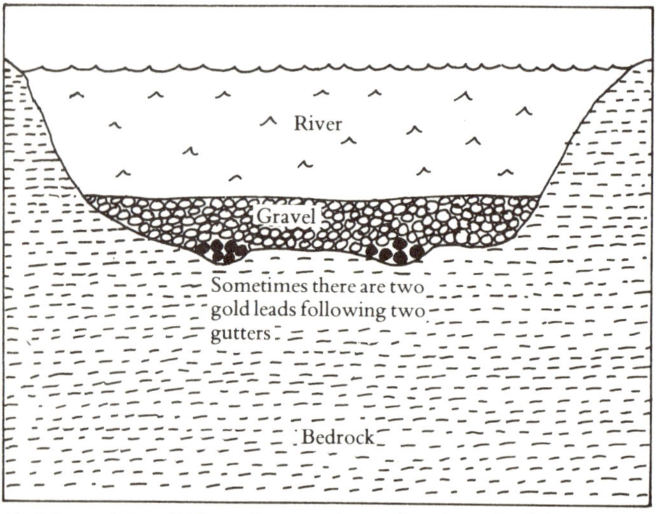


**Gold particles follow 'gutters' on the river bed known as leads. If there has been a heavy weathering of gold from the main reef it may follow two gutters or more.**

**The spot to pan in this river is in the shady area on the left. Because it is on the inside bend enough gravel has built up to trap the logs, and it may also have trapped gold. Notice how still the water is on the left. It would be impossible for gold to keep moving here.**

hillside? Isn't it enough to pick up just a couple of specks from the river bed in a week?' Those questions are valid. It *is* difficult to find a lead in the rivers because many have been worked over and over and over again by professionals who knew exactly what to look for, including second and third leads. But the point of describing the gold trail is to enable you to understand how it runs. You may decide to work on a river that hasn't been covered. Even if it has, gold may have been missed. The trail may run out as soon as you pick it up. But you should at least know what to look for and then if it *is* there you'll find it. So we'll continue to follow the trail through the dry gully.

This gully may fork at the foot of the hills and you will have to take samples from each 'finger' to trace your lead. There may even be a lead in each. Continuing our theoretical journey up the hillside you will notice that the gold is attached to quartz and other minerals because it has not travelled far enough from the source to become freed from these rocks. If the trail runs out you may have passed the main rock which has been releasing the gold. Or you may have taken a wrong turn. Without electronic assistance the only way you can trace your lost gold now is by 'loaming', that is, taking samples from several spots on the hillside, going up, down and sideways and panning each digging until you have traced the primary deposit.

Apart from twin leads, nature has tried a few other little tricks to keep her valuable inheritance from the prospector with the pan, but the old diggers worked them out after a while, too. Sometimes in the middle of a creek bed, with no leads to suggest it's there, is a whole pocket of gold grains. They have been trapped in a depression under the sand and then packed down by the sand itself, and the only way you'll find them is by pure luck or if you have been systematically working your way up river. They will have been trapped because the depression once had relatively sharp edges. 'Soft' depressions do not usually hold back the gold, particularly if the current is strong.

Other patches of gold may be lying from edge to edge of the creek bed, having been swept from the centre by a change of current brought about, perhaps, by the breaking away of part of the bank upstream. If you want to try to find these wider areas of gold you will have to do a bit of farming — digging portions of the river bed in designated sections. These divisions, not surprisingly, are known as 'paddocks'. You could square off, say, a width nearly

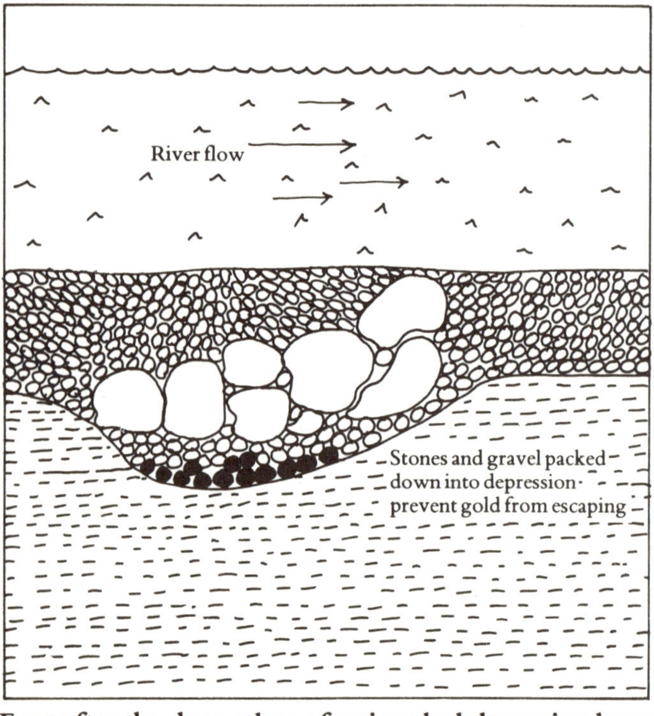


**Even after the sharp edges of a river-bed depression have been worn smooth, gold is sometimes still unable to escape into the current because of heavy rocks and gravels.**

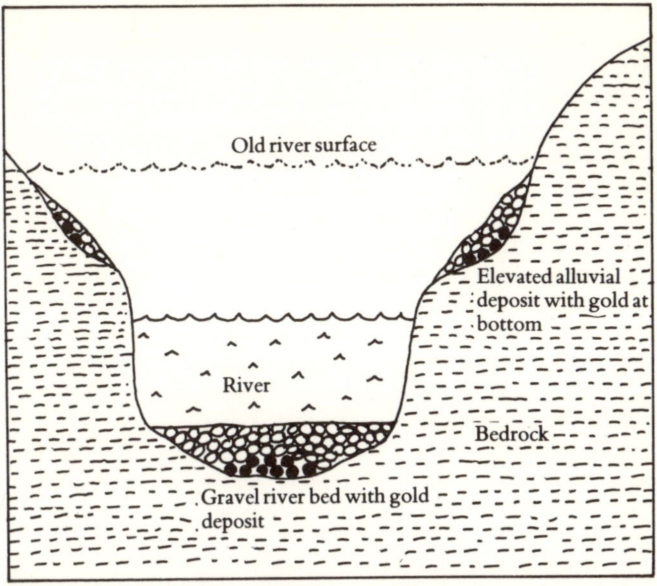


**When a new, fast-flowing river cuts deep into the bed of an old, slower-moving river, gold is left high and dry in the side gravels of the old bed.**

2 metres from bank to bank and thorougly dig and pan. The old prospectors did this quite often because it showed them how much work they had done for the amount of gold they had collected. It revealed later whether they had had a payable day. You may consider paddocking is too much work on a gamble or a hunch. It just depends how keen, or desperate, you are.

Another of nature's tricks which you should be constantly aware of is the *false bottom*. You dig through gravel and sand and reach clay. You assume the clay is sitting on bedrock so that is as far as you dig. But it could be that beneath the clay is another layer of sand with the gold underneath that. This is why you should dig as far as you can until you hit solid rock. If it's difficult to get the spade in, try loosening the soil with a stick or a pick if you have one, even if you have to stoop low and get the front of your shirt wet.

The panning prospector who is interested in the movement of water will realise that time has altered the course of many of Australia's rivers and creeks. Older rivers have widened, slowed, and have spread out across flood plains depositing gold over a wide area. Sometimes a wide, shallow water-course which contains a patch of gold from bank to bank may be considerably influenced by a surge of faster flowing water. The faster action of the water on the river bed deepens the channel, lowering the water level. This leaves the edges of the old water-course above the surface, and gold that is contained here is known as an elevated alluvial deposit.

Many rivers have dried up and the beds have become overgrown. Find one of these that has not been worked and you are on to something good. Water that brought gold down from the hills may have veered away from the original course, leaving the metal at the bottom of the dried-out creek bed. In popular gold-hunting areas you can be fairly sure that the main river-course, still flowing, has been worked out already many times. That is not to say that you will not find gold there because the metal can move on at any time. But the secret is to find the dried-out beds. A dried course may not have been found even in the mad gold-rush days and you may be fortunate to be the first prospector to work it. Finding an old course is really a matter of commonsense. Look around the lowest parts of a valley for any trough or long depression that could have formed an old river path. Unless it has been trapped in a crack or stopped by some obstruction further upriver, gold could be lying there.

The farther from the hills you search, the flatter and rounder the gold pieces are going to be; their journey down the river bed will have given them a battering. At the same time they will be purer because any silver alloy will have been partly removed.

If you have trouble locating a dried-up bed, go back to the main river or creek and look for any unusually straight stretches. These may suggest that a bend has been 'cut off' to form a billabong. If you can locate the cut-off section, that is a certain place to look. Clues are any curved depressions or trenches, particularly if they have a build-up of gravel at their sides. For the gravel to have been carried to that point, the spot must have been part of the river. The gravel may have built up because the old current was moving too slowly to drag it on, or because it was stopped by a dam of bigger stones. It is around these gravel deposits you should start digging. It will be hard work to push your spade in because you have no river water to assist you. On the other hand, there is no danger of the current robbing your spadefuls. If this is a very old river bed you will find the work particularly tiresome because you will have so much more topsoil and vegetation to work through before you strike the bedrock. Getting all the earth out is a backbreaking job and then you have to carry your dish of soil to water to wash it all through. However, dry panning can save you a lot of walking.

**Dry Panning**

*Equipment necessary:* One panning dish (preferably plastic with several riffles); shovel; pick; tweezers; wire; thick-bottomed boots; magnifying glass; tin can; glass jar.

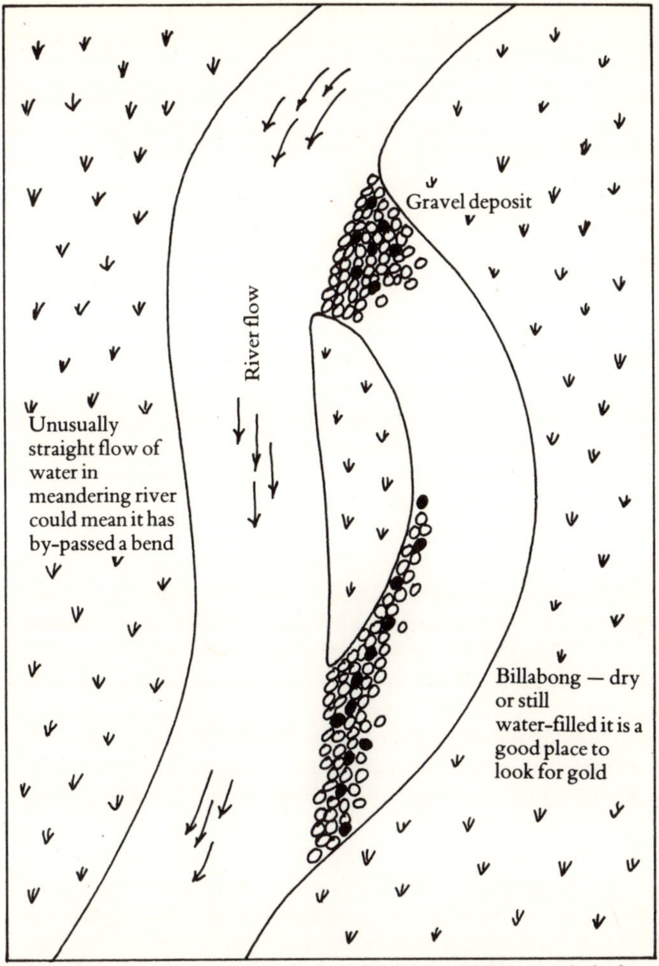


**As water runs out on a river bend when a billabong is being formed, the slowing current deposits gold.**

The equipment for dry panning is much the same as that needed for wet panning. In fact, you can continue to dry-pan with all your wet-panning gear, except that the pick, the thick-bottomed boots and the tin can are an asset.

Dry panning, it should be emphasised, is only a temporary method of sorting out the majority of stones and gravel when there is no water nearby. Water is still needed for the final, delicate stage. Many prospectors have criticised the system, claiming that it doesn't work. Perhaps it didn't in the days of smooth metal pans. However, the ultra-modern plastic pan has been proved to be successful for dry panning, providing you observe one very important rule: the gravel you work with must now be absolutely dry. Tacky sand is the prospector's nightmare if he has no water to break it up. It gathers up finer pieces of gold, and even if it is levelled out on a piece of paper the gold can still be lost between two or three grains of sand.

The ideal time for dry panning is in the dead of summer after a rainless period. Your material will then be very dry, unless you hit a damp area 30 centimetres or so down. The advantage of the plastic pan with its steep riffles is that you can take it a long way from water and still do most of your panning.

Suppose you have found a narrow, overgrown trench at the foot of a hill which you believe to be an old river-course. With your pick you dig down through the soil to establish the depth of the bedrock. If it is a reasonable distance down, loosen the soil further with the pick and then dig in with your spade, remembering, as in wet panning, that you must reach the hard base. Shovel your sample into the dish so that it is about half full, or less but not more than half full, otherwise you will not be able to create enough movement with any safety when you start rocking the dish. If the dish is packed too tightly with rocks, you will shake and tip with nothing happening and then suddenly the whole lot has shot out over the edge. If you aren't too sure of yourself at first, you could lay a piece of sheeting under the spot where you will be panning so that if you do accidentally spill soil at least you know where it is.

So you are now squatting or kneeling in some leafy glade with your dish half full of gravel and stones in front of you. There is no point in sitting at the side of the river bed in the blazing sun if there's no water to use. What you have to do now is push your hands into the dish contents, wriggling your fingers and generally mixing up all the stones. *Don't* put on rubber gloves to protect your hands even if you are a surgeon. There's a danger that fine particles of gold could stick to the rubber and you won't be able to feel them. Even if your bare hands become scraped and battered, you will still know if particles are sticking to them. Sift and stir, sift and stir, mixing everything. This process allows the small and heavy objects to drop to the bottom. You will also realise that this is another reason why you should not fill your dish by more than half. You need room for all this movement.

Now you can start to pick out the larger stones and bigger pieces of gravel, inspecting them carefully to ensure they aren't worth a few thousand dollars. Crack the big stones together to shake off any gold into the dish. If they have too much dust on them to distinguish colour, use a little spittle to shine an area with your fingers. Before you plunge your hands back into the dish, make sure there is no moisture left on your fingers. Sift some more, shake the dish, and now you can gradually start raking or brushing off more of the topsoil, ensuring that you do so across the riffles in the pan. The riffles are your safety valve and in dry panning you really need one. If you are careful and gentle, the gold will get caught up in them. Should you be using a metal dish with only one riffle you will have to be particularly careful (especially if you are a beginner), continually shaking your material back to the centre of the pan, tilting up again and brushing off, then shaking back.

When you are down to about two or three mugfuls of material you are ready for the next process of dry panning. Tilt the edge of the dish farthest from you downwards a little. You will be able to judge by how much as you work. If your dish has riffles — and the task is so much easier if it has — they should be on the side that is lowered, in other words away from you, tilted downwards.

Now, grasping the dish firmly with one hand, use your free hand to tap the upper edge of the pan. Do it carefully at first; you don't want to lose everything from the dish. Tap sharply several times because this will allow the gold to work its way through the material to the bottom of the pan and settle somewhere around the join where the base meets the side. You need to concentrate entirely on keeping the gold at the bottom of the dish and getting rid of the topsoil. As the larger pieces start to creep towards you, tilt the dish away even more and they will tumble forward out over the riffles. You must be careful that you don't tilt the dish too slightly, because the gold will tend to creep up towards you.

When you have tapped away most of the stones, bring the dish level and shake it and rock it so that everything settles back over the base. Then tilt and start tapping again, continuing the process until you are down to very fine sand and, hopefully, gold. This is where you must stop dry panning. Without the assistance of water it would be too risky to try to separate the sand from the gold, although you can blow away some of the lighter grains of worthless material if you are careful.

If you have elected to play safe and keep the remainder of the contents in your dish, give it a twirl as at the end of wet panning and try to spot any obvious pieces of gold. This is where your magnifying glass is useful. If you aren't experienced, don't trust your eyesight no matter how good it is. Obvious particles should be picked out and put in the glass jar. The remainder of the dish contents should be poured into the 'poverty pot' (an old coffee container would serve the purpose) for closer inspection with water either at the river or at home. Dry panning requires great care and it is not as effective as the wet method. However, it takes you more than three-quarters of the way to extracting gold particles and it does save you a long trudge to the nearest water source every time you bring up a spade of gravel.

# 7 Panning Country: Walhalla–Gaffney's Creek Goldfield

Time has marched past the deep valleys of the Walhalla–Gaffney's Creek strip of gold-bearing territory 120 kilometres east of Melbourne. The tall silvertop eucalyptus trees sway in the breezes that sweep across the rounded hilltops of the Great Dividing Range and whip up the fine dust that has been the curse of goldseekers for decades.

This is not metal-detector country. To find gold you have to go down as much as about a kilometre, or you pan it from the creeks if you know where the bedrock is not too deep. The old men who prospect in the valleys live on dreams of earlier days and it would be easier to squeeze blood from the stones they walk over than to learn where they find gold in the rocks.

They are wary of strangers, particularly 'new chums' with metal-detectors, and in the evenings when they gather together for a beer you can almost hear their mocking laughter echo across the hills. You can certainly hear it if you eavesdrop on their stories of the 'new chums' who have invaded the hills with metal-detectors they don't understand and with picks and shovels with the price tags still on them. They aren't just sceptical of metal-detectors — they are downright cynical.

It can be days before the old-timers acknowledge your presence and admit that they even go panning for gold. If they do you have achieved something but you will not learn any truths about where to go unless you happen to be a blood relative.

Jacky Reynolds, however, has broken the old prospectors' code. He has spilled the beans. If you want to find gold anywhere between Walhalla and Wood's Point you should heed his advice and 'leave the metal-detector at home. If you bring it up to this area it just shows how little you know about looking for gold.'

Jacky is sixty-five, although he feels ten years younger. 'I've been living the good country life, if you know what I mean,' he says with an artful chuckle on the verandah of the little weatherboard he shares with his brother opposite the old Post Office in

Walhalla. The house is full of rare old bottles that Jacky has picked up while fossicking around the valleys, and out the back is a battery — a rock crusher — where he pounds up quartz in the hope of extracting gold. He hasn't been making a fortune but he finds enough with his old steel pan to keep him happy.

'The trouble with this area is that the gold is a long way down. There is absolutely no point in bringing a metal-detector up here unless it can penetrate something like 4000 feet. I suppose if anyone could come up with a detector that could pinpoint gold at that depth he'd make himself a tidy little sum.

'Mind you, people bring detectors up here all the time. They've come up with some really "good" stuff too — nails, nuts and bolts, bottle tops, bits of wire and horseshoes. Why do they want to pay out $500 or $600 to find that stuff when I can go out and pick it up for nothing if I want to?'

Jacky, unlike some of his neighbours in other parts of the hills, has nothing against those who buy metal-detectors. But he says he feels a little sad for those who have bought a unit expecting it to work wonders for them. 'Their chances of finding gold in Walhalla with a metal-detector are far less than they are if they used a pan. Yet still they come up here, every weekend, a new group of hopefuls expecting to go away with a lump of gold in their pockets.'

He reveals that he does much of his local panning in Stringer's Creek, which runs parallel with the main street. His father, he recalls, panned 2053 grams (66 ounces) of gold in just thirty-six days from the creek, but Jacky doesn't reckon one man could pan as much as that now out of the short stretch of water beside the main street in thirty-six years. 'The old chums did a lot of work around here, let's not forget that,' he says. 'They were pretty desperate days and men laid down their lives to get gold. They put up with all sorts of hardships to find it and they weren't going to leave much in the creek beside the main street.'

Although Jacky admits that gold still works its way down from the hills into the rivers and creeks 'it's nothing like the amount that some people imagine. They think that after a good old rainfall gold is going to come roaring down from the hills. It doesn't work like that. Back in time, an outcrop containing gold would break away and all the gold would work its way down more or less together. So if someone in the old days found a bit they could usually follow the trail back and get all the bits. If anything is making its way into the rivers now it's very tiny particles, or there's a chance of finding a small bit that someone before has missed. They found a 20 ounce nugget in the creek once, but that was back in the old days. You won't find anything like that today.'

Jacky's advice to new chums who want to pan the rivers and

**Jacky Reynolds, 65, of Walhalla, Victoria. 'If you want to find gold, you've got to get down to the bedrock.'**

creeks in the Walhalla–Gaffney's Creek Goldfields — a 100 kilometre strip — is to 'get down to the bedrock'. He says he has seen a number of people scooping up sand when the bedrock is as much as 7 metres down. 'They haven't got the tiniest of hopes of finding gold if they stayed there for a year. Obviously if the bedrock is that far down you can't work at it. In Stringer's Creek it varies considerably. It can be 6 inches down or 25 feet down. If you don't know the area you have to experiment by sinking your spade in.'

However, Jacky imparts the information that the stretch of creek south of the Walhalla Hotel has bedrock as shallow as a few centimetres and can be easily worked. He won't swear there's gold there, although he suggests it is always worth a try.

So where do 'new chums' stand a reasonable chance of finding gold? Jacky smiles, does a bit of thinking and then says: 'All right, I'll tell you this much. You have to travel up a bit, north of Walhalla, to the Aberfeldy River and to Fulton's Creek. They are good alluvial areas. They got a lot of gold in their pans there in the early days and there's quite a bit left in the crevices in the bedrock.

'I know someone who has done quite a bit of searching for alluvial gold under water. He says that when he's in the Aberfeldy he can see gold lying in the crevices of the bedrock because water magnifies it.'

Gold lying in the Aberfeldy River, just waiting to be picked up? Is this the Australian prospector's El Dorado?

'Well you have to know *exactly* where to go,' Jacky slyly adds. 'I mean, you just don't go anywhere. Some spots are better than

**'See this thread of gold here?' says Walhalla storekeeper Phillip Turnbull. 'Metal-detectors cannot pick it up.'**

others. I'm just suggesting that the Aberfeldy is a good place to look generally.'

Jacky also suggests that some panning work in the Thomson River may turn up some colours, although the big rivers are difficult to work because the main gutter of gold will run up through the middle, which can mean cold and slippery work.

'In the old days they would try to follow the main gutter up through the river, looking for the main reef it was coming from. That's a very slow and tedious job and I don't think people nowadays would want to bother to that extent. When you start following the trail up over the hills you have to break open rocks to check if gold is in them and that you are on the right trail. It's a very difficult game and I think people would be just as happy to pan the small pieces they can find in the rivers.'

Jacky gathers up the pan and spade which live on his verandah and wanders down to the creek to do a bit of panning. What, he is asked, does he think of the new plastic pans?

'Don't like 'em. They don't feel right. But I suppose they're ideal for new chums because with all those riffles you don't stand a chance of losing any gold. Me, I'll stick to my metal dish.'

He uses it with remarkable speed. A shovel full of stones and gravel are reduced to a few grains in forty seconds. Jacky knows what he's doing. He also knows that this time he hasn't found a speck. 'I'll just keep trying,' he says. 'That's how it is when you look for gold. You just have to keep trying.'

Down at the local shop and tourist information centre, 25-year-old storekeeper Phillip Turnbull tells a disheartening story.

He has a piece of quartz with a thin thread of gold in it, about 5 centimetres long and easily visible.

'Everyone who comes in here with a metal-detector, and there have been a heck of a lot since the price of gold went up, wants to try his detector out on my quartz sample. Do you know there hasn't been one detector that has been able to pick it up.

'They have tried expensive models and cheap models. Oh, I think one of the most expensive showed a very, very faint response. One day I had a whole line of people in here with various makes of detectors and they were all saying things like "My model's better than yours, I'll get a response" but they couldn't get a beep. That doesn't say much for metal-detectors as far as I'm concerned.'

The road north to Wood's Point, where a 1462 gram (47 ounce) nugget was once found in the local creek, is a narrow, winding, dusty and at times treacherous, track. In parts it is just wide enough for a family sedan. Prospectors with trailers should stay away, for there are some very steep, unguarded drops. Although the residents of the tightly-knit community welcome visitors who are genuinely interested in the region's gold-mining history, they have been less enthusiastic about the new breed of goldseekers with metal-detectors who drive up from the cities in four-wheel-drive vehicles (or Toorak Taxis as the locals call them).

The first discovery of alluvial gold at Wood's Point, then known as Cherry's Point, was in 1861 and by August of that year some 3000 men were living on the upper reaches of the Goulburn River and hundreds of shanties were scattered through the hills. But what really put Wood's Point on the map was the discovery of the Royal Standard lode by W.J. Singleton and F.C.C. Stander. They were searching for strayed horses on a spur above the Black River, a tributary of the upper Goulburn, when they saw a huge quartz rock studded with gold. When they registered their discovery at Wood's Point there was a great rush to the area and claims were pegged out over a wide area. In 1864, as the mines began to open in a 30 kilometre radius, Wood's Point was a thriving community of more than 2000 people, with 216 weatherboard houses and stores, 114 log and slab houses and nearly 400 tents. Later the community dwindled following closure of the mines and the events of Black Friday, 13 January 1939, when bushfires sweeping up the valley destroyed the town. It has been rebuilt but the deserted main street suggests that some disaster has befallen. Strangers are conspicuous, particularly those with metal-detectors.

In the Commercial Hotel, whose atmosphere evokes the past, the barman is exceedingly polite and friendly to a newcomer who sits sipping a beer at the counter. However, in the corner a group of valley residents believing him to be a 'new chum', tell stories that would have any genuine detector-owner slinking out of town.

You drive on up the valley, through a virtual ghost town which is the A1 mining settlement, and reach Gaffney's Creek. A trained eye can see that all round there is good panning country: high hills, running water, slatey bedrock to catch gold particles. A pan man here is respected but, as in other parts of the valley, if you talk of metal-detectors you hear only arguments against them.

In the bar of the Gaffney's Creek Hotel, Ray Wright, covered in dust, is sipping beer with his mate Ian. They know what it's like to get gold the hard way. They are engaged in reopening the A1 Mine back up the road and if things go well they hope to have it operating within six months. Ray, the acting manager, is engaged in clearing the main tunnel and installing new support timbers, and you can see by the way he savours every sip of his beer what a relief it is to relax for a hour.

He says he's 'pretty choked off' with the advertising he has seen for metal-detectors. 'People who don't know anything at all about gold, who don't know where to look or even what it looks like, read these ads and they get sucked right in. They believe they can go out and find gold just like that. Sure, people have found nuggets. But what's the percentage?

'Know what I think? If you got a thousand people together who were planning to pay $600 or $700 each for a metal-detector and told them to invest the money in a gold mine instead, they are going to get a better return for their money. But no matter what method you try, you just cannot get gold that quickly. I've been trying for twenty years and I'm still working.

Ron Parker, who lives beside the Goulburn River at Kevington, says there is no doubt that gold can be found in the river but there are only two ways of getting it — by dredging and panning. A cradle can also be used to sort the material from the river bed. A friend of Ron has a portable dredging machine which drags sand from the river bed and he then puts it through a cradle. In one two-week period he amassed gold particles worth $2500.

'There's no doubt that the Goulburn is a good spot,' says Ron. 'It's hopeless territory for a metal-detector, of course, but people still optimistically pan the area and come up with gold. In my area alone I'd say on a weekend you can find someone nearly every hundred yards or so trying his luck.'

If you are a traditionalist with a pan, go to the Walhalla–Gaffney's Creek Goldfield. If you have a miniature dredger and a cradle, go to the goldfield. If you have a metal-detector, stay away.

You probably won't find any gold and you will almost certainly end up as a story in the local bar.

# 8 Metal-Detectors: What to Choose, How to Work Them

Since the boom in the sales of metal-detectors towards the end of the 1970s, scores of prospectors have made some very big gold discoveries. Without their detectors they would not have found the huge nuggets that now sit in bank vaults or are the subject of negotiation with wealthy buyers.

Some sceptics say that the chances of a metal-detector even paying for itself in gold are so slight they are not worth buying. But as they say with all lotteries, you have to be in it to win it; and you have a far greater chance of discovering gold than any of the thousands of diggers who invaded the goldfields in the 1800s.

You don't have to have a computer-operator's brain to operate a detector. An understanding of the controls is all you need, and sometimes you don't even need that. Detectors have worked so well for some amateurs that they have 'beeped' out the whereabouts of gold before the user could shout 'Strike!' One man found a nugget in a Coolgardie car park in Western Australia while he was trying to work out the controls and Philip Delecretaz earned the equivalent of a year's tax-free salary near Inglewood, 200 kilometres north-west of Melbourne, simply by switching on his detector while out on a barbecue trip with his wife and daughter. 'I had only just got going when the detector started to bleep very loudly,' he recalls. 'I started to dig and went down a foot before striking quartz. It contained about 567 grams of gold. I was dying to shout "Eureka!" but held back because there were other people nearby.'

If further evidence is needed of the fortune-changing success of metal-detectors, there is the story of the 30-year-old man from Melbourne's Templestowe suburb who lost his job and bought a metal-detector. The anonymous prospector went to a goldfield at Dunolly, 165 kilometres north-east of Melbourne, and, guided by the 'beeps', turned up one of the largest nuggets found in Victoria in recent years — a 1742 gram (56 ounce) chunk of gold lying 7 centimetres below the surface.

The Curator at the National Museum, Dr Bill Birch, made the remark after this discovery that many nuggets overlooked during the last half of the nineteenth century were being found with metal-detectors. The stories of success continue. In Western Australia a 74-year-old pensioner Bill Hay and his son Bob, while using a metal-detector, found a 3732 gram (120 ounce) nugget — 'Dad'n'Bob' — on a block of land 50 kilometres north of Kalgoorlie in the Broad Arrow Goldfield. The nugget was found about 30 centimetres below the surface. Another example is Ray Herridge, a 46-year-old machine operator from Kelso, near Bathurst in New South Wales, who found a nugget worth $7000 while using his detector in the hills near Bathurst. He had been prospecting for some time, mostly with a pan in creek beds, and had been using his metal-detector for only twelve months before finding the large nugget.

Metal-detecting does work for the gold-prospector. The finds described are those we *know* about. How many more dozens, hundreds, of big strikes have been made by modern prospectors who have wished to quietly and anonymously drop their big nugget into their sacks, put it on the mantelpiece at home and sit watching it every evening instead of switching on the television?

Obviously a working knowledge of your detector is half the battle; it's no good walking around listening to the 'beeping' if you don't know what its message is. Salesmen will demonstrate their efficiency by turning a piece of gold under the disc and the customer says 'oh yes, it makes a "beeping" sound when you do that'. It's not the salesman's fault if the customer doesn't ask questions. The reputable dealers will try to find out what you intend to use your detector for, but there are some fly-by-nights who will make the detector 'beep' and then wrap it up, knowing that you haven't got a clue what he is selling you.

How will your detector respond if there is a box of lead bullets lying buried between you and a nugget? How will it work when you pass over nails and pull-tabs from drink cans? Did you tell the salesman you wanted a detector only for finding gold? Are you buying a detector that will find silver coins, nicknacks and so on, but cannot find the most precious metal of all? The point of raising these questions is to emphasise the importance of buying the right detector. Some are really heavy on batteries. If rechargeable batteries are available, make sure this isn't because the unit eats through the voltage and leaves you without a 'beep' after half a day in the goldfields. Don't overlook checking the weight; pick it up and walk around the store for ten minutes — you may find it is too heavy for you. And make sure the unit has an authorised guarantee.

Detectors have developed considerably in recent years so that they can recognise, and not respond, to objects you are not interested in. By all means continue panning if you want to be nostalgic. But you will never be as rich as the man with the detector.

What type should you buy? You must appreciate that a metal-detector is a *metal*-detector. Don't expect to dig up a nugget every time the detector 'beeps', no matter how much you have paid. Despite an inbuilt 'brain' which enables many units to distinguish gold nuggets from nails, detectors can sometimes give confusing signals. They are only guides which indicate to you that something beneath the surface is worth investigating.

When you study the list of available detectors, the selection will probably confuse you — there are IBs, TRs, BFOs and VLFs to name a few. They may all seem like UFOs to you but this chapter will make it all clear. Prices range from a couple of hundred dollars to more than $700. Most have the ability to ignore unwanted articles like silver paper, nails and pull-tabs. Some are better than others for pinpointing old coins, but if you are seriously looking for gold you should be concerned only with a VLF-TR ground-cancelling model. If that has scared you off already, pause, and read on. All will be spelled out in the simplest of terms.

*VLF* stands for 'very low frequency'. Signals from a buried metallic object can be picked up from as deep as about a metre. Sometimes it is more, sometimes less, depending on the make-up of the ground, the size of the object and the diameter of the search-coil, the sensitive disc at the end of the detector's arm. The development of the VLF in recent years has made it possible for anyone with no knowledge of electronics or geology to find gold nuggets in ground that has a high mineral content with little or no interference to the signal.

*TR* stands for 'transmitter-receiver'. It is a sensitive 'brother' to VLF and helps the detector-operator to distinguish the make-up of an article beneath the ground without the necessity to dig.

*Ground-cancelling* has nothing to do with causing the earth to vanish beneath your feet so you can pick up all the nuggets for kilometres around. It means that the detector's brain can be tuned to ignore natural minerals in the ground, such as ironstone, which would otherwise affect its ability to give a continual reading when it is operated in highly mineralised areas. That was one of the problems with an earlier type of detector known as a 'beat frequency oscillator' and is, to a lesser extent, still a problem with single TR detectors.

Research carried out in the goldfields during the 1980 gold rush has revealed that many detector-operators did not know even these basic facts.

Often they were turning dials and flicking switches without understanding what they were doing. Consequently their chances of finding gold were considerably less than the prospector who had done his homework with the manufacturer's instructions. 'I was reading that everybody was going out finding gold with metal-detectors, so I bought one too,' said a Sydney man who, because of his obvious embarrassment over his ignorance of the controls, preferred to remain anonymous. 'I haven't a clue how to work the damn thing and I wish I hadn't bought it.' Asked if he had studied the instructions, he said he had 'scanned over them'.

Another man brought his detector back to Melbourne distributor Ron Mileham after a year and said: 'OK, I give up . . . how *do* you work it?'

In fact, if you are looking for gold only, the metal-detector can be a simple machine to operate because the surface areas of the Australian goldfields are littered generally with only two metals: iron to a great extent and gold, unfortunately to a lesser extent. VLF-TR detectors can ignore the iron and give you a signal over gold.

More than thirty models of detectors are available in Australia but as you are concerned with finding gold you should limit your interest to the dozen or so that work on the VLF-TR principle. That doesn't mean to say that a TR on its own has no use for finding gold. If the minerals in the ground are sparse, the TR is effective. The VLF-TR is a combination of two detectors which complement each other in the goldfields. If you were able to buy a VLF without the accompanying TR mode (single models of this type are not available), you would not have the use of the TR's ability to distinguish an old nail or gold. On the other hand, if you buy a TR only (and this type is available in Australia) you would not have the benefit of the VLF's ability to cancel interference from the natural minerals in the ground. It may be stating the obvious but when you have a combined VLF-TR, you use only one of the modes at a time — first the VLF and then TR. Side by side they are the most advanced system you can take to the goldfields.

Let's now look at the operation of one type of VLF–TR, a model considered by both amateur and professional goldseekers as one of the most efficient available. Obviously, different makes have different operating procedures, but they all work on a basic principle and the following guidelines, based on one particular model, can be applied to other makes.

The VLF–TR has a disc known as a search-coil into which is built two components know as the transmit-coil and the receive-coil. The transmit-coil sends a search message out to a metal object and the receive-coil collects a signal back at the 'target response area', which is where the two coils overlap. The message is transmitted through the search-coil cable up the stem, or arm, to the speaker in the control box which is close to the operator's hand.

The resulting 'beep' is synchronised with a dial — the louder the 'beep', the greater the response of the needle on the dial. The average search-coil is about 20 centimetres wide. Larger discs are available to seek out large objects at greater depths. A smaller search-coil is used to trace smaller objects at shallow depths. Many operators choose the smaller versions of 9 to 13 centimetres because these are lighter.

Before you can use the detector effectively on the goldfields, you must tune it to the ground conditions you will be operating over. If this is not done you will receive a variety of false signals that will finally cause you to take up panning. It is assumed you have read the manufacturer's operating instructions fully and familiarised yourself with the detector's controls. The following information is given in simple terms to help you overcome details that are unclear.

First, decide whether you wish to tune the detector with the automatic or the manual switch operating. Use of the automatic switch ensures that the detector will always keep itself tuned, whereas with the manual side of the unit working you may have to make minor adjustments from time to time.

Now you must find the detector's optimum tuning point, the 'pitch' at which it works best. Turn on the power and rotate the tuning control away from the null (quiet) zone until you can just hear a faint hum. Too much hum or not enough can result in unclear signals, rather like distortion on the radio when the tuning needle is slightly off the point of optimum reception.

The next step is to tune the detector so that it ignores the *mineralisation* of the goldfield, remembering that gold is a metal so you cannot 'tune it out' when you are using the VLF mode. Flip the master control switch to the left for one second; this activates the VLF ground-cancelling side of the detector. When this is set properly you will not receive a 'beep' and the needle in the meter will not respond as you sweep the detector over mineralised ground. To set the detector at this point of no reaction, lower the search-coil towards the ground. If the hum increases, raise the search-coil and turn the VLF ground cancel control towards the increase mark on the dial. Lower the disc again and, if the sound still decreases, turn the dial more towards the increase mark. Keep making these minor adjustments until the hum remains constant when the search-coil is lowered. If on the other hand the sound increases when the search-coil is brought down, turn the control knob slightly towards the decrease mark until there is a constant hum when the search-coil is lowered to between 5 and 10 centimetres above the ground. Make sure, of course, that during this process there is no metal close by, such as a pick or a trowel, that can interfere with the tuning.

The detector is now tuned to operate in the VLF mode. However, this system, on its own, has a drawback for the inexperienced operator.

You have tuned the detector to ignore the mineral make-up of the ground and you are now passing the disc over the surface. Suddenly, you hear a distinct 'beep'. The meter flicks to positive. You have found metal. What is it? The trained operator can tell from the type of 'beep' he hears what he is most likely to find if he digs. He knows that magnetic, iron objects produce a slightly wider response, or longer 'beep', than gold. The difference, in fact, is minimal and the ability to distinguish the signals comes only after many hours of practice in the field. However, how can the inexperienced operator assess his target without digging? He switches to the TR mode and it does it all for him.

The TR mode on some detectors is actually inscribed with headings, such as 'silver paper' or 'pull-tab'. When you receive a response on the VLF you switch first to, say, 'silver paper'. If there is now no response it means the detector has picked up a signal from silver paper and, by not giving a signal, you know that that is what it has rejected. If, however, you still receive a response, just as you did in the VLF mode, you then switch to the pull-tab marking on the controls. If there is no response you know that the detector has rejected a pull-tab. Should the 'beep' continue you have to dig the target because your tests on pull-tabs and silver paper have revealed that there is some other type of metal under the surface. It is only when the discriminating brain of the TR shows *no* response to items you are testing it against that you know those items are underground. It may sound a complicated, back-to-front system, but the machine is simply rejecting unwanted items by not responding to them.

The detector being used here as a basis for illustration does not have specific markers for 'pull-tabs', 'silver paper' or 'iron'. Instead, it has a discrimination dial numbered from 0 to 9. The lower the number the dial is set at, the fewer items will be rejected. If it is set at 0 it will not reject anything metal. By the time it is turned to number 3 it will have started to reject small iron items such as nails — in other words, if there's no 'beep' on number 3 you know you are over a nail.

But let's try some home-testing. First, press the master control switch to the right and hold it for one second to activate the discrimination control. Now lay the detector on a table and, with the discrimination dial set at 0, wave a variety of metal items across the base, about 2 centimetres away. You can use coins, nails, a belt-buckle, a watch, a pull-tab or a gold ring. By passing the items across the base you are simulating search conditions. At this 0 setting you are receiving a signal from everything. If you move the dial on to a low number, 2 or 3, you will find you are receiving no response from the small nails. If you turn the dial further you will find that other items start to be rejected. You may find that when you reach number 7 the detector is rejecting things like the pull-tab — but it is also rejecting the gold ring! It is entirely your decision at which number you want to stop the discrimination process. If you kept turning the dial on, to number 9, you would be rejecting the Welcome Stranger!

The goldseeker, however, does not need to concern himself with the higher range of numbers. On the goldfield he is likely to encounter, in the main, old nails and other small pieces of iron. These would be rejected at about numbers 2 or 3. It is impossible to say at what point gold would be rejected. So much depends on the size of the gold, the depth it is lying at, the make-up of the ground and the diameter of the search-coil. Some experts say it is safe to discriminate up to number 3 with no fear of cutting out a signal from gold. Other users, expert and novice, do not operate the discriminator at all. They contend that even a low discriminator number could result in the cancelling out of a small piece of gold. They rely on receiving a signal on VLF. The experts can tell if the signal means gold or rubbish, and the novice digs up everything.

Practice and a good knowledge of the controls of the detector are two paths towards success on the goldfields; there are others that will be discussed in Chapter 9. The operator must know his machine intimately, its advantages and its peculiarities.

For example, those with detectors with an automatic setting should remember that on some targets they will get an 'overshoot' reading. Sometimes the 'beep' will sound a split second after the needle has indicated negative. The needle will initially flick to a negative position, indicating that only unwanted material is beneath the surface, and then it will flick to a positive-response position. The man who really knows his machine and has done his homework will notice a sound reduction when he is over the target and will also be aware of a distinct hum on each side of the object as he moves the detector from side to side.

Experienced users are also familiar with 'hot rocks', which can confuse the deep-seeking VLF. These are stones of a very high mineral content. Because they are out of character to the general rock make-up of the area they can cause a 'metal' response on the detector. A localised 'hot rock' region can usually be determined by moving the detector around, and if it responds in a similar manner from nearby spots it can be assumed the mineralisation of the ground has changed — either that or you have found a whole nest of gold nuggets.

# 9 Finding Gold with the Metal-Detector: Where to Go, How to Do It

A nugget is officially described as a 'smooth, water-worn piece of gold that weighs more than one grain'. These are what you find with a metal-detector. Sixty per cent of your chances of picking up gold depend on going to the right area, thirty per cent on how you operate the detector and ten per cent on a combination of the machine's efficiency and luck. Research, then, is of prime importance.

You can start at a library by reading as much literature as you can on the gold districts in your state. Read the history of the towns, the stories of successes, the reasons why a town died. Did the diggings run out of gold? Did an epidemic cause the work to stop, thus leaving untouched gold? You should try to draw up a list of the type of prospecting that lies ahead — whether you will be scrambling around in holes or climbing over mounds of mullock heaps. An area that has been 'done over' thoroughly in the past may not appeal to you, and your research may lead to less disturbed parts. Perhaps you can establish from your reading whether there are good dredge tailing piles to work. These are the mounds of sandstone and shale dragged from a river bed by earlier miners who did not have the use of electronic equipment to thoroughly scan the piles. You should also be able to find out whether there are mullock heaps you can investigate — the mounds of waste material excavated from the ground at the old diggings. Perhaps, too, you can establish if there is a shallow river, within the scope of a metal detector, which has good placer areas, that is, sections where alluvial deposits of gold have become trapped. You may also be able to pinpoint dumps and tailing piles from old mines which can be a good source of gold (sometimes nuggets are concealed in rocks which have been discarded on the mine dump). Many detector-owners don't bother with research. They just drive to the nearest goldfield, switch on, and start scanning. There is nothing wrong with their approach because gold is where you find it. Your chances are improved considerably, however, if you bother with some basic research.

The Mines Department can also prove helpful. (Its name is different in some states.) Some departments have less information available than others, not because the staff have failed in their duty to the amateur prospector, but because their state may not have the same nugget-producing potential as others. An excellent item if you can get one is a list of the areas that have yielded nuggets. Such lists are not always readily available for mass supply, but most of the important finds are listed in Chapter 13. The Mines Department will certainly be able to supply you with a map that indicates the gold-bearing regions of your state. Perhaps as the hunt for gold continues, Mines Departments will make nugget lists generally available, although some lists will be very short. South Australia, for example, has not been a major gold-producer and its Department of Mines and Energy has a record of only six 'large' nuggets, ranging from 373 grams (12 ounces) to 2488 grams (80 ounces). Victoria, on the other hand, has an outdated list of 230 nuggets that were lying in spots so shallow they could have been found with a metal-detector, although a large number of deeper nuggets is also recorded.

Obviously the best places to look for nuggets are those that have yielded the highest number. In Victoria, the Dunolly region has produced 126, made up of one over 31 kilograms (1000 ounces), four over 15 kilograms (500 ounces), 62 over 3110 grams (100 ounces), and 59 over 1555 grams (50 ounces). The lowest recordings are attributed to Chewton, Jordan, Marong, Middle Creek, Newstead, Pitfield and Warrandyte, each of which has produced only one nugget over 3110 grams (100 ounces). It would seem, therefore, that based on these records, Dunolly is the place for the detector-owner. You are no doubt wondering: 'Have all the nuggets now been gathered from Dunolly? Are there another fifty nuggets waiting to be picked up in Chewton, Jordan, Marong . . . ?' Your research cannot tell you that with any certainty, although the history of the area you have selected will give you some indication of your chances. If, for example, you found that Pitfield was scorned by early diggers because it was proven to be virtually barren and that the nugget found there had dropped off someone's cart, your chances in the area can be said to be low.

Further and very worthwhile research for the metal-detector owner is into the depth at which nuggets were found.

Unfortunately, the records at all Mines Departments do not show such details but history books in state and local libraries may reveal the information. Local museums which proudly display replicas of large nuggets from the district will almost certainly have a record of their depth. In Victoria, which must be referred to again because it has produced Australia's largest number of nuggets, the Department of Minerals and Energy sells a photocopy of an out-of-print publication listing all nuggets over 622 grams (20 ounces) found in the state. The publication, *Memoir No. 12 of the Geological Survey of Victoria* by E.J. Dunn, also records the depth at which the nuggets were found. These details are a valuable guide. If you found that all the eighteen nuggets found at Rokewood were discovered at depths of about 3 to 4 metres, you may assume that is the depth of the auriferous wash and any other rich rocks in the area will be beyond the capabilities of your detector. You can be sure, though, that there are thousands of nuggets lying at shallow depths in *or near* the old goldfields of Australia waiting to receive a friendly message from the search-coil of a detector. Many will have been actually dug up by early prospectors who would not have noticed them in all the dirt, and the nugget's weight would not have made a detectable difference to the weight of the pile on the spade. The unnoticed excavated nuggets would have been dumped into the mullock pile. In fact when the 31 kilogram (1008 ounce) Heron Nugget was found at Fryers Creek, Victoria, by two prospectors, they noticed it had pick marks in it from earlier workings. Some very good regions to visit with the detector should encourage you with their name alone: Nuggety Gully (Vic.); Cutter's Luck (W.A.); Golden Point (S.A.).

You will make your nugget-hunting easier if you decide to work as the diggers of old did — in pairs, or threes. There is nothing more tedious for a single operator — he could still be classified as a 'hatter' — than to have to keep putting aside the detector, gathering up the pick, digging, rechecking with the detector, digging a little more, rechecking . . . Apart from that, the activity is more enjoyable if a family or several people can be involved. As one metal-detector dealer in Dandenong, Victoria, said: 'Going into the country with the family and a detector is the next best thing to a sausage barbecue.'

When you have decided from your research on the area to search, you will probably be surprised at your first visit to a goldfield because, as you approach, it will look so much like the surrounding countryside. If you have decided to search out old diggings, you will have to walk into the trees before you spot evidence of all that has gone on 130 years before. And then the evidence will surround you — hole after hole after hole. You will also notice the mullock heaps, the piles of earth shovelled up, carted to a nearby spot and dumped. They may even be beside some holes. The area will resemble a battlefield and it could be said that is what it was, while man struggled to win gold from hard-packed ground that he hated like an enemy.

First, tune your detector to the mineral make-up of the ground. The most likely area to start is that part of the goldfield that has yielded the most. Your research may have already told you that. If not, establish from local residents where the main encampment was because that was usually set up close to the rich areas. If you cannot get this information do not worry, because if you work the goldfield thoroughly you will pass through the best area anyway. You may wish to start by jumping into a hole. But have a look first. Some are only about a metre deep because that is as low as a chum sometimes went, and other holes will have been filled in to that level. However, other holes are quite deep and sometimes bracken can disguise the actual depth, so take great care.

At the more popular diggings you can assume that someone has been in the hole before you with a metal-detector. But this does not suggest that any nuggets that may have been in the hole have been traced. Many were missed by the inexperienced 'new chums' in the 1850s and you can be sure that there are many modern novices with metal-detectors. For a start, the novice may not have known how to operate his detector properly. He may have failed to have tuned it to the correct ground balance and received a signal almost every time he moved the search-coil. He has dug and found nothing. He has received another signal and repeated the negative search. Every time he has moved the search-coil up or down he has received a signal and, frustrated by his efforts, has left the hole for the next fellow.

Even if your predecessor had been more familiar with the workings of his detector it is still possible he overlooked a nugget in his haste to search out as many holes as possible. He may have been a wizard with the controls but he has not known how to search properly. The two things must go together. A hasty searcher misses nuggets for two reasons: he is scanning so quickly that he overlooks a faint signal from a deep-lying nugget; or he has not brought his search-coil the necessary 2 to 5 centimetres above the ground when he comes to a depression, therefore lengthening the distance between the nugget and the search-coil. Both have the effect of weakening the signal or cancelling it out altogether. The only way to gain all possible depth is to maintain a constant distance between the ground surface and the coil. So do not assume that the 'obvious' holes (those close to the road or to a parking area) have been thoroughly scanned. To use a detector properly requires patience. A number of inexperienced searchers will claim that not only have they covered several holes in one day but also several *goldfields*. However charming they may be, don't listen to their advice. In fact, give them some of your own by gently advising that searching so quickly is like not searching at all. When you hear some experts claiming they know every rock on a goldfield, their exaggeration is very slight. What they are saying is they've done a very thorough job. And it's so much more satisfying to settle back at the end of the day with a beer knowing that

**Old digger's hole at Moliagul. Don't risk going into these, especially when the earth is cracked around the top, as it is here.**

even though you didn't find gold you did the job properly and thoroughly, at least as thoroughly as the detector allowed. What you as a detector-operator have to remember is that although you can work faster than the man with a shovel you cannot always work as thoroughly. He *has* to dig deep to search for his gold. You switch a button and listen for a 'beep'. If a nugget is about a metre or more into the ground you may well miss the signal. Meanwhile the man with the spade keeps digging and an hour after you have wandered off he chances upon the nugget your detector did not pick up. At least that's what you thought at the time; in fact the detector did respond, very slightly, but you in your haste to move on missed the signal. This comparison between you and the digger is made to emphasise the necessity of working very patiently.

You can cover much more *surface* ground than the digger because your detector can cope well with shallow depths. But for nuggets deep down you must move slowly in order not to miss a signal. As you don't know the depth nuggets are lying at you should be slow and thorough at all times. The experts scan nearly a

**Old diggings trenches should be checked thoroughly. Do the sides as well as the bottoms. Don't worry if there are signs that others have been here before you. They may not have been very thorough.**

metre each side of their feet and move forward the diameter of the search-coil every one and a half seconds.

Around the diggings where there was little water and much of the separation of earth from gold had to be by dry panning or dry blowing, your chances of finding nuggets are very good. Much was missed in the earlier days because equipment was so crude. The nuggets may not be lying on the surface, but many will be less than 30 centimetres or so down, having been covered by dirt excavated by the diggers themselves or by natural processes over the years. This is why it is so important to properly adjust the ground control on your detector and to maintain a constant distance of about 5 centimetres between search-coil and ground surface. The search-coil operating height will always be determined by the amount of mineralisation in the ground and sometimes it may be necessary to increase the distance to as much as 10 centimetres.

The mullock heaps are among the best spots to search — those piles of discarded material at the sides of the holes. New chums made a lot of mistakes and often threw aside valuable nuggets. When gold was dug out with the wash-dirt it was hard to distinguish from stones. In some areas, too, gold was coated with iron oxide, which also made identification difficult. The valuable

**Make a point of scanning 'edges' like the side of this shallow trench. The old-timer who made this trench may have missed a nugget by a fraction.**

wash-dirt often became mixed with the worthless mullock, particularly when it was left piled up near the mullock while the prospectors waited for the rains that would clean the dirt through.

Nuggets were also overlooked when the diggers worked late into the evening. As they operated their cradles in the poor light, they sometimes missed the precious metal they were looking for. Whenever you see a pile of rubble or stones at the diggings you should investigate. You should be prepared to level some of the piles — those keenly into detecting call this process 'hammering the heaps' — so you can get an even sweep with your detector over every part of the rubble. Prospectors who went over the old diggings in the 1920s were convinced they were on to a good thing because they were aware that in the 1800s nobody could detect what was in the ground. Of course in the 1920s they were not able to see what was below the surface either, unless they carried out extensive digging. They, too, will have missed much of what they were looking for.

If you're in a gold-bearing region with dried-up creeks and river beds you should search literally every part of these. The chum who has worked the water-courses before you may have scanned only an odd metre here or there, missing big patches. By patient scrutiny you have a chance to collect what he missed. But first, you must establish the depth of the bedrock. If you're still digging at 3 metres without hitting anything hard, it might be an idea to move downriver where the current would have been slower and the water-course shallower. This doesn't always follow but is worth remembering. If the bedrock is between 2 centimetres and about a metre down you should be able to detect lumps of gold. Your hopes of tracing flour gold or scattered specks are dim. But let's not be too critical when the detector misses the finer stuff; it's more than enough, surely, that you can track and unearth nuggets you couldn't see.

Having established that the bedrock is within penetration range along, say, a 45 metre stretch of dry water-course — digging your pick in is the simplest way of finding out — you should now look for any natural line of stones coming out from the bank which may have been a 'bar'.

It's impossible for a nugget to come to a stop on an *unimpeded* stretch of the water-course unless the river was suddenly snap-frozen. Even if the water had been dammed and started to slow, causing the nugget to lose its suspension, it would have struggled slowly on downriver until something, perhaps quite small, caused it to stop completely. It's being assumed that you are looking for nuggets and not scattered specks or grains of gold which your detector would be hard-pressed to find. These small pieces may well be lying on a seemingly uninterrupted stretch of river because they are small enough to become caught up in tiny crevices in the bedrock. As a nugget-hunter, though, you should be looking for the more apparent obstacles that could have held up a larger piece of gold.

It may have been suggested to you that a smaller search-coil is easier to use among the bigger rocks of a dried-out water-course. That's perfectly true. It is lighter for a start and it enables you to get in between rocks where a bigger coil will not fit. However, you will find more gold if you keep to the standard-size coil, not only because it gives you a slight edge on depth but *because it makes you work at finding gold.* You come, say, to two rocks that are 15 centimetres apart. A small coil would squeeze in but not the standard type. With the larger coil you either have to forget that spot, or move the rocks. If you can move the rocks, move them. Not only will you then be able to scan the gap, but also cover the ground on which the rocks were standing. If you were using the small coil you would be satisfied you had got between the rocks and there was no need to move them, and then miss the area on which they're sitting. You pass the coil over the top of them. And if they are 30 centimetres high, that's 30 centimetres you have lost. There are no short cuts to gold-hunting with a metal-detector. Perhaps one day someone will design one on wheels, like a golf cart; but you will still have to get out and move the search-coil

**This is where your dreams could come true: at the mullock heap. These are the piles of earth and stones thrown up by the old diggers in their quest for gold. Many nuggets were tossed into the mullock heaps by mistake, as present-day hunters with their metal-detectors are finding out to their delight.**

among a cluster of rocks in the same way as you have to get out of the cart to hit the golf ball. The machine can only help you up to a point. So if you use the average-size coil you will be forced to move things. The more you move *off* the surface, the greater penetration you have *into* it.

Of course there are some things you can't move, such as mountains. If the bedrock curves up from the creek bed to form the bank, you will probably find it has large cracks in it. When the water was flowing it would have created an excellent gold trap if the level of the stream was high enough and the current fast enough to keep the gold moving with the gravel on the river bed. If the cracks are on the inside, slow bend there is even more reason to believe gold could have become caught in them, This is where the small search-coil is necessary because you will never be able to widen the gap to take your 20 centimetre coil (unless you use explosives and

**The roots of old trees have been known to trap gold. You may also get false signals caused by gaps under the earth created by the roots as in this picture. If you aren't sure what your detector is telling you, dig and find out. Remember the detector is only a guide and the final decision, to dig or not to dig, is up to you.**

for them you *must* have a permit). A crack full of gravel strongly indicates that the water level did once cover it; but just as importantly it suggests that the hole has not been worked before. Perhaps it has been examined but the searcher did not take the trouble to clear out the gravel. Either that or he was a very tidy worker and, having removed and examined the gravel, piled it all back in the hole again. When following in the steps of a man who is meticulous about prospecting etiquette, the inexperienced goldseeker has difficulty deciding whether someone's been there before him.

A certain type of gold trap is found around the dried-out creeks of the diggings: the roots of trees that protrude from the bank. They will be partly covered with loose stones and shale which became caught up when the river was flowing. It may not be possible to get your search-coil in among all the roots, so dig and scrape the gravel away and carry it to a spot where you can test it electronically. Then go back to the roots and try them again as best

**A perfect nugget trap. When this old creek was flowing, gold could have become caught in this gap in the bedrock. Those small, dusty stones that lie there may not be worthless. If you can't get the search-coil right into the gap, scrape the gravel out and examine it.**

you can. One of the biggest nuggets found in recent times, the 6718 gram (216 ounce) Father's Day nugget, was caught in a tree-root so these are worth thoroughly investigating (see Chapter 3).

As you move along the creek bed, watch for areas where the bedrock protrudes above the stones and gravel. Any crevice between two raised portions of bedrock should be carefully checked and if you cannot get the search-coil in between them you should take up your trusty coil once more and examine all the scrapings you can get from the gap.

If you are following a wide water-course, don't keep your head glued to the ground all the time; look up occasionally around the nearby hillside for evidence of a much higher water level. You may ask: 'How can there be any nuggets up there? Surely, as the water level subsided with a deepening of the channel, gold will have kept to the lowest point?' As a rule, that's true. But what you're looking for on the hill is an elevated alluvial deposit which, as explained earlier, is formed when a river changes nature. A wide river may have dried up and hundreds or thousands of years later started flowing again with a narrower, faster-flowing current. The current will have cut a channel in the middle of the old bed and lowered the level of the water, leaving the sides of the old course high and dry.

At all times you should be confident that your detector is working efficiently for you, even if at times you receive false signals. A goldseeker who lacks confidence in his own ability, or in that of the detector, lessens his chances of finding gold. He doubts that he is going to be successful so he doesn't investigate thoroughly; he doubts his detector will find him any gold so he doesn't pay much attention to signals unless they sound like the Tumbarumba Brass Band. From time to time you will receive confusing signals, but you should not let these deter you. Every model of metal-detector has its peculiarities and as these cannot be altered you must learn what they are. With experience you will undoubtedly receive many false alarms because your detector will pick up signals from unexpected objects. It's generally fair to assume that most of the objects lying buried will be nails, horseshoes, scraps of iron and other items that the discriminator indicates aren't worth digging for. Then you receive a good signal that is not cancelled out, so you assume, then, that it is not coming from a scrap of iron, and begin to get excited. Confident that it can't be anything modern, you scrape back the topsoil and find a pull-tab. This example is an unusual one but the more the diggings are scoured by modern prospectors the more this is going to happen. By the turn of the century, metal-detector operators will be turning up articles ranging from a nineteenth-century bullet to a twentieth-century keyring — unless by then a detector has been developed that will ignore absolutely everything except gold.

One of the false signals you could receive at the diggings will be caused not by 'hot rocks' or the rejects of modern living but by clay. You will find this occasionally when you're working in one of the holes which is fairly deep. Clay deposits can play havoc with your equipment by giving a fairly faint metallic indication which suggests a deep nugget. The problem is caused by neutral clay, which contains neither mineral nor metal. Your detector has been tuned to the surrounding terrain. You have adjusted it so that when passing over the generally mineralised ground it does not respond. However, when you passed over the clay deposit, the detector's 'brain' accepted a different signal and passed it through to the response needle on the control box.

There is another problem with clay; deposits that have some mineral substance in them and should have been cancelled out can give a faint reading if the earth contains too much moisture. When you dig out the clay and test it with the detector you find that it responds positively. The signal will not be as strong as a solid nugget, but it may be enough to cause you to begin the time-consuming process of breaking the clay up into small pieces. You may even take it down to the local creek and 'puddle' it, expecting to find a piece of gold in your panning dish. You can save yourself the trouble, though, by recognising that as you break up the clay when you have first excavated it you are getting a signal from each separate piece.

In rare circumstances conductive clay can cause you to miss a nugget. If gold is lying fairly deep and close to a deposit of conductive clay it may send out the same strength of signal as the clay. So you dig out kilograms of clay, discover it is conductive, and assume that the other signal close by — the 'message' from the nugget — is also from clay. These circumstances are unusual, but they emphasise the point that you should be as thorough in your digging as your patience will allow. By adopting a thorough attitude you will learn a great deal about the make-up of the soil and how it affects the response in your detector. You will learn that everything that responds as metallic is not necessarily metal; that it just happens to be conductive. Iron pyrite, 'new chum's gold', gives a metallic response although it is classified as a mineral.

You are now aware of factors that can cause the detector to give a false reading. Pockets of uncharacteristic minerals are the most common and will create a respond signal because you have tuned the detector to a different mineralisation. At the diggings in the north-west of Victoria, operators have reported yet another cause of interference: hollows in the ground formed by the removal of boulders or tree-stumps. The detector will respond with an audible increase because the distance between the ground and the search-coil has lengthened.

When you receive a signal that prompts you to investigate, great care must be taken when you are digging. Your partner, if you have one, should carefully brush away the surface rubble in an area about twice as big as the search-coil. Because the VLF has poor pinpointing capabilities you must expect to have to dig patiently. When the surface rubble has been cleared, test again with the detector. If the signal has stopped you know that your target is in the excavated soil, so it is wise when first brushing it back to keep it in a neat pile. Some prospectors like to use a sieve on their dirt piles but there is little advantage in that unless you are smart enough to be able to see that among the big, dirty stones is a nugget. By using a sieve all you are doing is making more work for yourself because, apart from the sieving operation, you are presenting yourself with two piles to test — one of big stones and one of little stones. With a metal-detector you will isolate them anyway during the scanning procedure, as will soon be explained.

First, go back to the hole you are digging. After digging away a little of the topsoil you are still getting a signal from the spot where it was first received. Start now to dig a flattish hole with a small pick or a trowel, putting the soil from this in a pile *separate* from the first pile. Otherwise, if the signal has stopped from within the hole and you have put your second pile on top of the first, you will now have to investigate the whole lot. Each time you clear 2 centimetres or so of soil, investigate the hole again and put the latest waste in a new pile. It doesn't matter if you have six or seven little piles around your feet; keep digging and adding more to the collection. The further down you go, the louder the signal should be.

At all times you must exercise great care with your digging implement. If your pick strikes hard against a beautiful heart-shaped nugget, the damage would reduce the value as a collector's item by thousands of dollars. Some detector-operators keep digging and digging and then go back over the soil they have excavated. A lot of work is saved if you examine the soil as it is dug out. When at last you have stopped receiving a signal from the hole you can assume that the object is in the last pile of dirt, the other piles having been checked. If you cannot readily see the cause of the signal, spread the dirt from the last pile on top of the search-coil and remove it a bit at a time until the signal stops. You then know that the stone you have just taken away and are holding in your hand is — well, good luck.

The diggings are an excellent place to begin at and indeed to stay around. You may, however, wish to trace other sources of gold. Wherever you search in a recognised goldfield, it has to be there and you may want to attempt to trace a reef deposit in the 'mother lode' in higher ground. Just as you start off by tracing the thin thread of gold leads in streams and creeks, so you can follow the clues up through the gulleys towards the original deposit. Don't expect this to be easy. Although it is known that large amounts of quartz are a good indication that gold reefs or lodes are nearby, the quartz itself can sometimes be difficult to trace. Often the quartz outcrops are obscured by soil or vegetation. However, if you can find quartz or particles of gold in dry river beds or gulleys you can expect that in the nearby hills there are gold reefs.

Geologists have little doubt that nuggets and finer alluvial gold originated in auriferous quartz reefs because many nuggets have had quartz attached to them. The prospector with a pan has a greater chance of finding fine gold in the river bed than the metal-detector operator, but the detector-user has the advantage of being able to turn up 'the big one'. He is also likely to trace the source of the valley gold far quicker than the pan-prospector who must patiently trace the tiny leads up through the river to the bank and from there up through the hillside. However, if the metal-detector operator can trace a hillside 'float' he has a chance of following the trail to the primary deposit.

Extracting large amounts of gold from the primary deposit would be difficult to say the least for the casual prospector, who, under the rules of his Miner's Right, can use only hand implements. All types of gold originate as an ore vein created by volcanic activity millions of years ago and some veins run deep into the hillside. But if you can trace an exposed reef you may be able to chip pieces out and have the quartz and the gold crushed in order to extract the gold; or you could keep your chippings as a reminder of your toil on the hillside.

Even if a prospector can only get close to a primary source he may find float gold in abundance. A float, you will recall, is ore that has broken away from the main outcrop and can be found around the main body of rock containing gold. Water, gravity and wind carry the floats down the hill and they become lodged in hollow spots or behind big rocks and other barriers. The floats, which can range from large boulder-size to fist-size chunks, sometimes drop into pockets on hillsides to be covered by an overburden of rock and gravel that has been washed in over the top. The metal-detector is invaluable for tracing these pockets as long as the overburden is not too thick and the section of rock bearing the gold is not on the underside. The bigger the search-coil in these circumstances the better because of the advantage of added penetration.

The search for floats should begin at the bottom of the hill, or where you consider that rocks tumbling down by gravity or in heavy rainfall have stopped. Quite often it's just commonsense to work out their resting place. If you can find a pile of very large boulders with smaller rocks beside them it is worth investigating with the metal-detector because the smaller samples may turn out to be floats. Should you get a positive detector-response, try to break open the rock and examine it. If they are gold-bearing floats, you will be able to determine from their position the direction they have travelled down the hill. Try other rocks as you move up the hill, remembering that the ore is a combination of mineral and metal so the signal on the detector will not be as strong as with a solid metal object. Faint signals should not be overlooked, particularly on a flattish section of the hill. This is because you may be standing over a hidden pocket which is covered by other debris brought down by rains.

While on the hill you may also detect an 'ore chimney', the signal from which could be much the same as that from a pocket filled by floats. The chimney is a much deeper finger of metallic and mineral ore which has been forced up through cracks and fissures in the host rock by volcanic pressure.

Apart from finding traces of gold on the way up the hill, and you will be more than happy to do this, another indication that you are on the right track towards the primary deposit is a change in the rock structure. Not even a brilliant geologist would safely say that a change in the rock make-up means a primary deposit of gold is nearby, although it is an encouraging sign. If when you were searching lower you were surrounded by shale and now you can see quartz veins in the rock face, you may be close to gold. If loose rocks nearby are sharp and angular that is another indication that you are close to the host rock that has shed the floats. You should also watch for whole outcrops of quartz, or areas where rusty-looking quartz is in the hillside. All these indications are good, but if you are still unable to pinpoint a vein of gold it is time to start a grid search.

Tune the VLF mode of the detector to the surrounding ground (at this higher level the mineralisation may have changed) and search in lines, say 15 metres at a time, scanning a fairly wide path with the search-coil as you go. If you haven't received a signal at the end of the first stretch, turn and walk back to your starting point in a parallel line to the first path but 2 to 3 metres from it. You will possibly find it helpful to make occasional marks in the soil, or build a small pile of stones, to indicate the line of your first path. Should the detector 'beep' increase, you may be over a vein or a pocket of ore. To be sure you have a correct reading (you don't want to hack into rock if you're mistaken), walk back a few steps and the sound should decrease. As you walk up to the spot again, the 'beep' should resume. If it does, clear away the surface soil and try to establish the cause of the signal. It is virtually impossible to see gold in veins unless a strong magnifying glass is used. If you think you have found gold, you should be able to trace a length of the vein with the detector. You can use a hammer and chisel to try to get gold out but it is not a simple task. However, if you succeed in getting a quartz sample which, based on the signals from your detector, contains gold, do no more at this stage than put it in your pocket.

In addition to use on the diggings, on hillsides and dried river beds, a VLF–TR detector can be operated in water to trace nuggets. Don't expect to get signals from small particles of gold, especially if they are scattered through the crevices of bedrock. The detector won't pick them up. The unit will have to be adjusted to ignore the negative effect of the highly mineralised creek bed and to overcome a temperature change that would affect its performance. The search-coil is then operated about 5 or 7 centimetres from the surface of the bed. Metal-detector operators have been amazed at the amount of man-made junk they have found in the most isolated parts of a river. Nuts and bolts have an annoying trend of sinking into the sand — and exciting your detector. All the time, though, you are eliminating rubbish and the next spadeful you bring up may contain what you are looking for. You should also use your detector while at the riverside to scan over the dredge tailings of earlier prospectors. It is possible that a nugget was mistaken for a stone and dumped.

Yet another good search area is at the site of a disused gold mine, providing you seek permission if the property is fenced off. Many old-timers used to wet the rocks to see if gold stood out and, if not, the extracted lumps were discarded. Piles of dumped rock can be an excellent target for the metal-detector operator, although you should remember that a number of 'tailings' may be no more than debris from tunnels that were dug to get to a rich vein. It is important, then, to test *all* the tailings by examining rock samples from each pile. This can be tedious work because the barren regions around the mining areas are not as pleasing visually as the uncut hillsides above alluvial digging areas. It requires some mental effort to sift through dusty rock piles after you have experienced the relative ease of using a detector in a small river or creek where the water has sorted out the rocks and the gravel and the gold to some extent. In addition there are a greater number of clues around the river: you now know that gold can become trapped by boulders, that it sinks to the bedrock, that it stops travelling when the current slows. On mine tailings, gold could be anywhere or nowhere.

**Golden Rules for the Metal-Detector Owner**

1 Don't ask 'Where is the gold?' You may receive some answers you don't expect.
2 When you dig a hole, fill it in behind you.
3 Don't dump waste batteries around the goldfields.
4 When a sign says 'Keep Out', do what it says.
5 Respect a pegged claim and keep off — for your own well-being.
6 Don't crowd the other fellow. Until a claim is pegged, Crown Land is open to everybody, but give others at least 10 metres breathing space.
7 Don't disturb people who aren't looking for gold by working the metal-detector in their picnic spot.

# 10 Metal-Detecting Country: Dunolly District

Old Jack Flett thumps his fist down on the table in the tiny museum at Dunolly in Victoria, and exclaims: 'Metal-detectors — bah! I wouldn't give twopence for one.

'There's only one way to find gold and that's with these' . . . and he raises his gnarled hands high. 'They found big nuggets with their bare hands in the old days around these parts and they used their commonsense. Most of the people who use metal-detectors don't know anything about gold. There's no skill required. They just wave them about and wait for a noise. That's not gold-prospecting. That's cheating.'

Jack is quietly reminded that cheating or not, a number of people have been making some money by turning up gold lately, here in the heart of Australia's nugget country.

'But they've also *wasted* money in buying them things,' he insists. 'I don't need a metal-detector to go out and find gold. I can do it in the good old-fashioned way — with my eyes and hands.'

Jack has a very good argument when the 'old methods' are questioned. He claims that six years ago he picked up off the ground a 124 gram (4 ounce) nugget a short distance from the town centre. He looks you straight in the eye and says: 'Ah, you don't believe me. Well, I'll show you then!'

And he reaches into his back pocket and brings out a small glass bottle with a blue top sealed down with sticky tape. Inside is a piece of cotton wool and sitting on the cotton wool is a jagged nugget.

'What about that then, eh?' he asks as you take the bottle which has been thrust at you. 'That's what you call a nugget, isn't it? A real little beauty. And I picked it up because I used my commonsense.'

Jack is eighty-four. Born and bred in Dunolly, he worked his way around Victoria before returning to his old home town twenty-one years ago to run the intriguing little museum. He is there at the door every day collecting 60 cents a time from tourists and the owners of metal-detectors who, after a frustrating time out in the surrounding goldfields, come into the museum to gaze in wonder at the one that got away — a replica of the Welcome Stranger which was found about 16 kilometres away.

Jack has earned himself something of a reputation for collecting an extra dollar a time from detector-owners who want to test their units on his nugget. He doesn't miss many tricks, old Jack, and what he doesn't know about gold is not worth knowing anyway. His younger brother worked in a local mine and his cousin James Flett was the author of a book about the history of the Victorian goldfields. Jack reckons he knows where a lot of the gold is lying around Dunolly way and although he is happy to give away a few of his secrets there are one or two things that he says he would rather keep to himself. This is quite understandable of an octogenarian who has the enviable ability to pick nuggets off the ground.

'It came to me as a premonition,' says Jack, tapping the bottle with the nugget and then laying it on the table so that all visitors to the museum on this afternoon can gaze upon it. 'It had been raining pretty heavily and I had this premonition that I was going to pick up a "slug". Well, the first day I went out and did a bit of panning and didn't find anything. You see, I knew that after the heavy rains there was a chance that gold would be washed clear somewhere. After that first day I hadn't given up hope and the next day at seven o'clock in the morning I went to one of the diggings a mile south of Dunolly on the Maryborough road. One of the new chums had been working in one of the holes, oh about 112 years ago, and he had gone 8 feet down straight through the wash-dirt and he was looking in the clay. See, he didn't know what he was doing. He had brought up a bucketful of earth on the way down and the nugget was in it. Well, when I went around, there it was right before my eyes all nice and clean. Funny, though, because I half-expected to find something. Like I said, I had a premonition about it.'

Jack was buying gold at the museum up to 1979. People were bringing in the odd ounce or two before the detector boom began. But when the price of gold started to go up he closed his till. 'Wasn't going to take any chances on buying gold at a high price and then finding it crashing all around me. Wish I had bought now, the way it's gone up. Still, I'm happy enough with my nugget.' He hasn't computed its gold value for a while but he knows its somewhere between $2500 and $3000.

Jack, whose real name is John, is the last old-timer left in Dunolly who knows everything about gold. 'I'm afraid they've all gone now. Perhaps it's just as well they don't know what's been happening out on the goldfields lately. All those gadgets sounding off. You won't find anybody who'll dig a hole and jump into it nowadays. They'll scratch down a couple of feet and then off they'll go again.'

He says that old diggers had a lot more patience and enthusiasm and he's talking back to 1856 when it was estimated there were 30 000 to 50 000 men in the Old Lead and when the main street was nearly 5 kilometres long. The lead was mostly nuggets and the wash-dirt in one spot was said to be thick with gold. Two hardworking diggers are reported to have brought a hundred-weight of gold from one hole in five days. Dunolly is a mere hamlet against the bustling community it was then when there was no room to park a cart in the main street and where wild scenes went on late into the night. A report later in one newspaper recalled:

Dancing saloons there were in abundance, and all night long the jingling of out-of-tune pianos, the scraping of villainous violins and the fitful blowing of cracked cornets, fell upon the savage breast of the otherwise civilised person, robbed of his sleep. Then there were the concert halls. Our shop was situated next to one of these. Songs of more than questionable morality were bawled out nightly, but one never became inured to the sights and sounds of this den of infamy.

Jack missed all that, of course, and although he spent much of his life away from Dunolly he returned occasionally to 'keep in touch'. By doing so he learned a lot from the old-timers who had links with those earlier days and with the mines. He knows that the area is still rich with gold, despite the incredible amounts unearthed in the rush.

'Oh, there's no doubt about that. You have to have a lot of patience, though, to find the small stuff. The bigger stuff's easier to find. I have been out and washed twelve or thirteen dishes with no colour and then on the next one I have got a few bits. Mind you, you have to know your area and . . .'

Wait, Jack. What does he mean when he says the 'bigger stuff's easier to find'? How easy?

'Oh . . much easier. I know where there's heaps of gold around here and all I want is someone to come along and invest in it.'

A little prompting, and Jack imparts some of his secrets. The heaps of gold he talks of, so it transpires, are deep underground 'but you can get to it easily enough'. He continues:

'There's the Queen's Birthday Mine, for a start. It was back, oh, when was it, somewhere in the 1890s when there were twenty-five men working there. They were hauling out rich ore and they knew the company was making a fortune. Well, those men decided they weren't going to work their hearts out for a pittance so they went to the boss and asked for a rise of a shilling a day. The company had made its money by then and they weren't at all concerned by this demand. Know what they did? They closed the mine, that's what they did. It never reopened. And do you know what? I've got the working plans for it. Trouble is, though, you wouldn't get men to stick at it today. They'd start with $200 a week and the week after they'd want $300 and on it would go.'

Jack pauses a moment before deciding to go on. 'I'll tell you something else, too. There's the Bromley Mine. My younger brother George worked in that. It was the last mine to operate in this area. There's gold in there all right because George told me all about it. It's been closed for twenty-five years now but it should never have been allowed to.

'What happened was that the company went down deep and came to a seam of gold in the rock. Well, they wanted to follow the seam under the creek but they had to get the Mines Department in to inspect it first. The Mines Department took one look and refused permission for them to go any further. They said it was too dangerous to go under the creek because it would all collapse. That mining company sold out and a new company came along and sank a shaft down on the *other* side of the creek to try to pick up the seam. They brought up a few samples of stone and put it through the battery and found they had bitten right into the gold seam. They didn't stay long. They worked out their costs, found they had made a nice tidy $20 000 profit and closed the mine. They left an awful lot of gold behind, believe me. It's out there now, 300 feet down under Burnt Creek. I don't reckon its dangerous. If I had the equipment I'd have a go at getting to it.'

Jack, who is supposed to be taking admission fees from tourists, has an audience at the entrance to the museum. The incoming tourists, holding their 60 cents admission, seem rather mystified by this elderly gentleman standing here telling everyone where they can find gold.

'There's another reef I could put you on to,' he continues. 'I reckon you'd get 6 ounces of gold to the ton out of it. When you reckon that in the 1880s an ounce to the ton was considered payable, 6 ounces at today's prices isn't bad, is it? Mind you, you'd have to install the right machinery which would cost something in the region of $30 000 but you'd soon get that back. I don't know why the mining companies aren't flocking out to this area. Never mind your metal-detector people. But perhaps the companies are concerned, like I was last year, that the price of gold will crash and they'll lose their money.

'You know, there's a creek out the back here that runs alongside the railway line. The old-timers reckoned they got all the alluvial gold out of it, but I don't know. I reckon there's a lot of gold left lying about if you've got the right hunches and the patience to look. If I get another premonition I'll be out there right enough.'

Jack's suggestion for the present-day prospectors 'if they will insist on using metal-detectors' is to go over the mullock heaps around the diggings. 'The fellows that were there 120 years ago did all the work so somebody might as well try to benefit from it. The new chums made a lot of mistakes and there will be a lot of nuggets that they missed lying around out there.'

Outside Jack's museum is a rusting anvil mounted on a cairn. The inscription beneath tells visitors that it was on this anvil that the Welcome Stranger nugget was broken up for weighing. Jack is very proud that it was he who arranged for the cairn to be built and the anvil to be displayed and you can see him exercising self-control when, on being offered a pamphlet by Jack about the world's biggest nugget, a lady tourist asks: 'What's the Welcome Stranger?'

She has not been to the monument up the dusty track through the gums at Moliagul which marks the spot where the 78 kilogram (2520 ounce) nugget was found just below the surface by two penniless Cornishmen. Metal-detector owners head for this area first because the Dunolly and Moliagul goldfields have also yielded at least twenty other nice nuggets from shallow ground. On any weekend now scores of modern prospectors scour the old diggings in among the trees, and Hazel and John McCoy, who run Moliagul's quaint general store and petrol pump, couldn't be happier about the newcomers. In 1855, 4000 men were digging at Moliagul and the area was dotted with tents, but until the boom in metal-detectors McCoy's Store had seen some quiet trading days.

'I've never known it to be as busy as it is nowadays,' says Hazel during a break from serving customers. 'I've been here fifteen years and there's never been anything like it. There's no doubt about it that everyone's got it into their heads they can find gold. It's been a big boost to our little store. I've even had to open up on Saturdays and Sundays which is a thing I'd never done. This area is the spot that everyone heads for. I suppose everyone is looking for another Welcome Stranger nugget, but to be quite honest I haven't heard of anybody finding anything.'

Says John: 'It's almost as if the past has come alive again. There haven't been any prospectors of note in this district for years but now here they are in their hundreds. Most seem to have metal-detectors. They seem to be standard equipment nowadays.'

Up on the hill near the Welcome Stranger monument Cleve Hill from Maryborough is actually digging up the main track. His wife, Sylvia, has just picked up a response on her VLF–TR and Cleve is using a pick to hack into the track surface. The pick he is using is of interest because he designed and made it himself specifically for prospectors with metal-detectors.

Engineer Cleve found after he had purchased his detector that there wasn't a great choice of hoe-picks so he turned out what he considers is the 'perfect design' and sells it for $16.50. Now, Sylvia is waving the detector over the hole in the track while Cleve digs away with his special pick. They work as an experienced team. Sylvia moves in. A 'beep' from the detector. Sylvia moves back. Cleve, in grey overalls, moves in with the pick and digs again. Cleve moves back. Sylvia moves in. Another 'beep'. In moves Cleve. They come at last to some blackened earth.

'It looks to me as though someone has had a fire going here once,' observes Cleve. 'That suggests to me I'll find a nail that was in the wood.' Some more 'beeping' and digging and the target flicks out of the hole, having been caught by the point of the pick. Cleve picks it up. 'See,' he says triumphantly, 'a nail!'

'But have you found any gold?' he is asked.

'Gold . . . ? Oh no. No gold. But I've heard of someone who has been making a heck of a lot of money. He's a man in Deniliquin who deals in metal-detectors. Seriously, in *Deniliquin!* I'm told by someone in the business that he has been flogging scores of them. Now, I asked myself, what has Deniliquin got to offer to the gold-hunter? No gold in Deniliquin. And then I found out. Everybody in New South Wales is coming to the Victorian goldfields. And a lot of them pass through Deniliquin, don't they? I tell you, if you want to make a bob or two, set up a metal-detector stall on one of the Victorian borders.'

The Hills have not brought their detector out without having researched the best areas and bought maps from the Department of Minerals and Energy.

'I wonder if we're ever going to find any gold,' says Cleve. 'We think we're doing all the right things. We read the instructions that came with the detector very carefully. It responds to metal well. We know that because we keep digging it up. Unfortunately it's always worthless. We have, however, got this long-term plan. We're going to invest in salt. The old miners worked so hard in the mines that we're going to go in there and gather up all the salt they sweated out.'

Cleve is happy with the way his specially-designed pick works and, across in the trees, Manfred Hohenhaus is happy with his specially-designed metal-detector. Yes, he made it himself. In fact he has made one for his wife, Pamela, one for Anita, 13, one for Karen, 11, and another for Tania, 9.

Manfred has used some quite basic parts. He holds the Hohenhaus detector by a rubber grip designed for a bicycle. The main handle is off an old broomstick. The plywood search-coil is held on by part of an old squeegy mop. The sensitive parts come from a variety of electrical stores, but he had to wait a few months before he could get an indicator dial. 'The shops were sold out,' he says. 'I think everybody has been trying to make their own metal-detectors.'

Manfred, a delivery driver from the Melbourne suburb of Rosanna, read books on how to make a detector and he learned more at an electronics class. Five wires run from the search-coil to the control box and the whole thing runs off a nine-volt battery. The total cost was $40.

'It works well,' he says enthusiastically. 'Look,' and he drops a bunch of keys on the ground. He lowers the wooden search-coil, lower and lower. Nothing. 'I've cancelled out too much,' he says, explaining that in order to cut out the interference from the ground minerals he has turned down the response volume. He turns it up as he holds the disc over the bare ground and this time there is a crackling.

'That's the ground minerals,' he says, and then he moves the disc over the keys. The gadget 'beeps' loudly. Well, it finds keys all right, but has it found gold?

'I haven't had it long enough yet,' says Manfred. 'This is the first time I have used it. I think when it is over a nugget it will respond. Yes, I'm sure of that.'

Stan and Ethel Pfeiffer from Swan Hill are operating their TR detector beside a mullock heap. They have bought the higher frequency model because they are gem collectors and find it is more suitable. Today, though, they are looking for gold and like everyone else who is within about a half kilometre radius of the Welcome Stranger monument, they haven't found any. They are working their fourth goldfield after unsuccessfully trying around Maryborough, Bendigo and Maldon.

'I studied up on the detector and practised with it for a while,' says Stan. 'I follow the gauge, not the sound. The more the gauge flicks, the more solid the metal. It's easy for *me* to understand *it*. But I don't know if *it* understands *gold.'*

Detectors, though, are proving to be confusing for some people. Ron and Rita Harris have travelled from Sydney with their VLF–TR and Ron admits he hasn't 'got the hang of it yet'. He says he was told by the dealer that he could be sure of finding gold but after three days and scores of nails he is a little unhappy about the sales talk.

'It might be me, though,' says Ron, a 33-year-old painter. 'I don't know what the discriminator does. I read the instructions but I now think I didn't spend enough time on them. Well, actually, I didn't read them. Rita read them out to me in the car as we drove down. I don't know where the instructions are now. Might be in the back of the car somewhere.'

Wattle Flat is tucked in among the trees beside the Moliagul–Dunolly road and from here comes the distinct sound of an interrupted 'beep'. Here a former Collingwood footballer, 38-year-old Neville Withers and his father, Cecil, aged sixty-six, are fighting with their VLF–TR. Neville is being terribly good-humoured but it is surprising his detector isn't wrapped around the nearest tree. However, he is a sportsman so he is being very tolerant.

'I was talking to a chap this morning,' he says, 'and he had a piece of quartz with gold in it. So he says "See if you can pick that up on the detector" and I put the detector over it and it didn't register. I don't know what's wrong with it — perhaps I haven't got the tuning right. It seems to be going off all over the place and when we dig there's nothing there.'

Sure enough, as he spoke, the detector 'beeped'. Cecil dug and found nothing. 'That's what's been happening all the time,' says Neville. 'But I'm picking up a few little tricks here and there by talking to people. For instance if you've got steel caps in your shoes they can interfere with the detector reading. And you mustn't get your pick in the way. Apparently that's quite a big mistake a lot of people have been making. You dig a hole and lay your pick down beside it and you get a reading. You dig further and lay your pick down beside it again and get another signal. You could spend days making that mistake until you caught on.'

In among these trees at Wattle Flat, though, serious goldseekers are at work, such a 29-year-old welder Bill Woodman with his VLF-TR complete with arm-rest. He is accompanied by some determined assistants — his mother, Mrs Patricia Woodman and his girlfriend Janine Arthur, who are carrying plastic panning dishes and trowels.

Bill, from Noble Park, Melbourne, is working his way along a narrow, dried-out creek, moving slowly, scanning from side to side.

'I did a lot of research,' he says. 'I spent several weeks reading all the books I could get hold of on modern goldhunting. I must admit they weren't very clear. I had to work a lot of things out for myself but I was determined to do everything possible to learn. I got maps from the Mines Department and read about the goldfields. When I went to buy a metal-detector I bought the best I could. I spent a long time studying it because no matter what people tell you about them the only way to work them properly is to practise with one yourself and get to know just how it feels and how it sounds.

'Yes, I understand the theory of the detector . . . how VLF penetrates deeper than TR but TR is better for pinpointing objects. I know that gold-bearing country is mineralised and that you have to tune the detector to cancel out that mineralisation in the ground in order to receive a clear signal from gold, which is metallic. I know that when I am searching I should keep to a certain routine so I know what areas I have covered — like, I know that I have made a thorough job of this creek because I have started in a particular spot and I know that between here and there I have covered it from bank to bank. I know that it's safer to dig up every item that responds on the VLF because it could be a nugget or a nail. I know you should pay particular attention to faint signals because it could mean a deep-lying nugget. There are a whole lot of other things I know about searching and about the detector and yet despite all this I still haven't found a single speck of gold.'

Some 2 kilometres from the Welcome Stranger memorial is the modest weatherboard home of 42-year-old Arthur Deason, grandson of John Deason, one of the two prospectors who found the famous specimen. Arthur laughs as he holds up a tiny flake of gold. 'This is the best I could do,' he says.

'People seem to think it's a mystery what John did with his money. I know. He went broke before he died. He sank $1000 that the family knows of into a mine which failed. My father worked for him and in a way I have my father to thank for the tiny flake of gold I have now. I found it by his front gate after the workmen had carted up a lot of stones to make up the road.

'I haven't got a metal-detector, not me, but my brother Dick has bought one. I suppose some people might think it amusing that descendants of one of the men who found the Welcome Stranger are resorting to these gadgets. But I'm strictly a panner myself. I know where you can get flour gold, the really tiny stuff. In fact, I used to tell people they could use my paddock to look for some but I had to put a stop to it after one fellow dammed up the creek and left it like it. Most of the people that call at this house asking for advice are pretty good, though. They don't know who I am but as this is one of the few houses around here, they'll knock on the door and ask where the best places to go are. Most of them are pretty good fellows. If I can put them right, I will.'

It is late afternoon now in Dunolly township. The museum is about to close down. Jack Flett has put his nugget back in his pocket. Cars with metal-detectors in the boot are heading back to the city. Walking up the main street towards the milk bar is Manfred Hohenhaus, with his family. There are no signs of the Hohenhaus broomstick detector.

'Find any gold?' he is asked.

'Aw . . . it got too hot,' he says with a smile and a shrug. 'Yes, too hot. Maybe another time.'

# 11 Equipment: Ours & Theirs

In the early days of the Australian rushes, when shallow alluvial gold was more abundant, it was either collected into bags for cradling or panned at the river. Recovery methods gradually became more systematic and efficient, and modern equipment has made the searching and recovery work a relatively simple task. All you need to know is where to look but the manufacturers of the modern machines cannot tell you that. When you have finished this book you will have a good idea of the best type of places to search: as for equipment, obviously the more money you are prepared to spend, the more you can buy and the faster you can work. There is still no guarantee you will become richer than the man downstream with his pan. For the patterns are the same as yesterday. Two men working a rocker at the diggings could work faster than the chum at the creek but they had no more guarantee of finding gold than the modern chum has with his dredge. The gold game is a lottery. They said it before at the original diggings and they are saying it in the goldfields that have been revived with the sound of metal-detectors, and at the rivers that are being sucked by dredges. You either hit the spot or you don't.

The point of the modern equipment, of course, is to save toil and it certainly does that.

Perhaps the greatest advance in working alluvial deposits in rivers has been the development of the dredge. If there is to be a division in the rapidly expanding world of gold-prospecting, it must surely come between advocates of the metal-detector and proponents of the dredge. The detector-operators will hold aloft their nuggets, and the dredge-workers will retaliate by pointing out that a huge percentage of gold recovered in the world has been in such fine particles that electronic equipment could not have traced it.

The old-time prospectors would have been more than happy with either system. They certainly had nothing to match the metal-detector but they did have wing-damming. Worked in conjunction with a crude pumping system, wing-damming blocked off a section of a river so that prospectors could pump out that area and get to the bedrock where gold would be lying. The modern dredge needs no supporting dam. It simply sucks up everything on the river bed while the current flows on above.

Working a wing-dam took a good deal of organising. Damming rivers is no longer allowed, but the efficiency of the dredge makes this unnecessary. The old prospectors would drive a number of stakes into the river bed in a semi-circle from the bank and it reached out as far as the middle of the river. The gaps between the piles were blocked off and made watertight by stuffing them with wood, clay and stones, forming a pond-like area of water which could then be pumped. A similar water-isolating technique was applied where the flow of water was weak. Straight dams of clay and wood were built out from the bank anything between 9 to 36 metres apart and stretching about three-quarters of the way across the river. The water continued to flow past the unfinished end of the dams, and the area between the blockades, now no longer being fed by the current, was either left to dry out or was pumped. Pumping in both damming systems was usually carried out with the aid of a California Pump, a continuous belt passing over rollers with buckets attached.

The buckets passed into the isolated water when the belt was operated either by man-power or water-power, and they then travelled to an elevated part of the pump where they were tipped and the contents flowed away. Finally, the river bed was exposed for working. Even in modern times mountain prospectors have built their own pumps, using tins instead of buckets and getting their power from a river which turned a small water-wheel.

But here is the dredge, the efficient river-bed vacuum-cleaner which can chop many hours off older bed-scanning methods. It is only in the past two or three years that dredges have become popular in Australia, and the serious prospector regards them as the best means of recovering gold from river areas and creeks that were almost impossible to work during the gold-rush days. Nooks and crannies in the bedrock that could be reached by hand, spade and brush were never cleaned out entirely but they can be reached and cleared by the probing nozzle of a dredge.

Nevertheless, dredging can be a drudgery and can be expensive. The cheapest models on the Australian market cost about $1000. For that, you get a 6 centimetre suction pipe, a pump and engine, a sluice box and a plastic float unit on which the whole apparatus sits. Proponents of the dredge insist that what you pay out in cash you get back in gold and then more. Their argument is a good one and they don't have to use it because the dredge fraternity aren't anxious for gold-hunting hordes to join them on their relatively quiet patch of river. Certainly one unchallengeable argument is that the dredge enables the modern goldseeker to explore areas that have never been checked before. Even if river sections have been wing-dammed and pumped by earlier prospectors, the chances of their having found every speck of gold in every cranny of the bedrock are extremely slight. When the dredge is employed it sucks the crevices dry, figuratively.

Australia has a range of some twenty models of dredges with suction hose diameters ranging from 6 to 20 centimetres. The measurement is taken from the top end of the hose because the 'working' end (the nozzle) is slightly smaller to block off stones that could cause a jam-up further along the system. The smaller version is capable of drawing up a third of a cubic metre of loose bed gravel an hour and the 20 centimetre model can cope with 2½ cubic metres.

As dredging is becoming increasingly popular in Australia despite official limitations, it is worth looking further into this method of prospecting.

The workings of the dredge are quite straightforward. It sucks up everything, rejects unwanted material and retains the gold. At least, that's the idea. The suction hose is connected to a vacuum device known as the eductor. A powerful vacuum is created when water is pumped into the eductor under high pressure and the vacuum is employed to draw up gravel and sand from the river bed. This collection of stone and particles is first washed into a baffle box where, because of the turbulence created, clay concentrations are broken up. The dredgings are then forced up against a screen mesh. Only the smaller pieces pass through the mesh for sorting against the riffles on the tray of the sluice box. The larger waste stones which have been accepted through the mesh, fly quickly over the riffles and are ejected at the discharge end. Meanwhile the smaller particles, those that will contain gold, are carried up to the riffles in a relatively slower current of water that ensures that gold is not kept suspended. The lighter material is carried out over the riffles. Gold and heavy particles are trapped and settle onto a matting designed to stop escape. The riffle section is then lifted away and any gold and sand can be removed easily from the matting for final panning in the faithful dish.

The dredge-operator searches every part of the river that the experienced panner looks at and more. Large areas of rivers in Australia have been untounched by goldseekers simply because they could not get at the river bottom. Even the Chinese, remembered with respect by old-timers because of their efficient gold-collecting methods, were helpless to clear out the crevices in the river.

'I'm surprised they didn't come up with some sort of dredge,' says 35-year-old John Wilson from Melbourne who is building his

**Two types of dredges, one with the sluice box attached; the other with a detachable box which stays with the operator. The detachable sluice box is lying on the far side of the stream.**

own machine. 'They were amazing at clearing up all the gold in their path but they didn't get it all out of the rivers. That means there's still a lot of gold left in them.'

In fact, John Morris, a 39-year-old goldseeker from Gladstone Park, Victoria, estimates that the average end-of-year profits for individual dredge-operators are higher than those gained by prospectors with metal-detectors.

'I'm not criticising metal-detectors — I use one myself,' he says. 'But I also use a dredge and I have dealings with a lot of people who go in for dredging. Therefore I know the amounts of gold that people are finding. You see, it doesn't make the newspapers when a dredge-operator gathers 5 ounces of gold in a week. Now the man who picks up a 50 ounce nugget with a metal-detector gets publicity but that's it — that's all he finds in that year or in five years. Meanwhile the man with the dredge gets another 5 ounces in his second week, and another 5 ounces in his third week. Relatively speaking, he is doing a lot better in a year than the man who finds a single nugget.'

John Morris has no doubts that dredging is quickly becoming more popular in Australia. In the limited stretches of water that are legally open to dredging (Chapter 13) he saw an increase in the number of dredge-operators in the early months of 1980. He recalls one day when he spotted thirty dredges 'jostling for space' on the 7 kilometre stretch of Goulburn River between Jamieson and

Kevington in Victoria's Great Dividing Range. However, he makes the point that that is one of the few spots where dredging is legally allowed.

'Australia is a great country for the dredge-operator at the moment because it is such a relatively new event in prospecting here,' he says. 'It's been around for many years in the American goldfields but it hasn't caught on in a vast way in Australia — yet. Not long ago an American drove into the caravan park near the Goulburn with a dredge on the roof of his car. In ten days he had collected 28 ounces from the river. What he was doing was driving around Australia dredging the rivers. He must have been wondering why there was so little competition.'

The old panners not only had to be content with working in shallow areas or building wing-dams; they must have realised they were not doing a thorough job on the river bed. The modern dredge-operator does not have the same worries. He can either work the narrow rivers by standing in his wellington boots beside his floating dredge; or he can put on a wet-suit and dive in the deeper parts, leading his suction hose to the river bed.

A weekend prospector, considering the purchase of a dredge, should look no further than the 6 centimetre model, which is easy to operate for the man who wants to stay in shallow water. Recovery of gold depends, of course, on the operator 's knowledge of where to look, but the actual workings of the dredge are simple.

'If you can start a lawn mower you can start a dredge,' says Leigh Moore, the managing director of a Melbourne-based supplier of prospecting equipment. The smaller dredges, weighing about 53 kilograms, are described by Mr Moore as 'the perfect compromise between portability and high production'.

Despite the relative lightness of the dredge, to be a successful operator you have to be 'utterly dedicated and physically strong', according to John Morris. Often, he points out, there are many big stones to shift and, having moved them, you find the bedrock is hard and polished and certainly not gold-bearing territory, so you have to patiently move on and start again.

Before heading for the river bank with the dredge you first need to investigate the spots you want to operate at. Obviously this depends on how close you can get the car to the river. You don't want to spend half the afternoon scaring the 'roos from the bush as you stagger through their territory with the dredge on your back while you look for water. You will need to spend some time searching out a point of easy access.

The water need be no deeper than 20 to 22 centimetres and, unless there are obvious gold traps, which will be outlined shortly, you should start in the middle of the river. Newcomers to dredging usually make the same mistake when they have got the dredge operating. They start working on the upstream side of the area they intend to work, and then find as they move down that their dredge tailings have all been thrown into the river path they are moving along. That means they now have to dredge up the stuff they have already rejected in order to get at unprocessed gravel underneath. You may notice a similarity here with the metal-detector operator who puts a new pile of dirt on an old one and then has to scan the whole lot. Unfortunately, with dredging, about 60 centimetres may be worked before the 'new chum' realises he should have started *downstream* and worked *up*. By doing it this way, he not only makes less work for himself but he also fills in the holes he has created; the rejected material pours out behind him to make a neat patching-up job on the river bed.

In dredging, as in panning, commonsense often comes first, followed by experience. Once you know the theory of the movement of placer gold — how it will bury itself in any hiding place when the current is unable to keep it travelling — you can often work out for yourself where it could be resting. At times you don't even have to think about it because your growing experience will tell you which areas to try. You may have already tried to clear some nooks and crannies below the river surface with wire and picks before purchasing a dredge. You may have had a feeling that you left a lot of good stuff behind.

The upstream side of boulders and rocks that form a barrier, or bar, across the river are the first places to try with the dredge nozzle for two reasons. If there's any gold around, this is a certain spot for it to build up; and if you have difficulty remembering procedure the rocks will force you to work your way upstream, preventing the mistake of throwing all the tailings out in your path. The river bed, under the overburden, can be a veritable honeycomb of gold. If the bedrock is slatey or creviced, particles will have been nestling in the niches for thousands of years, and you are now living in an age when equipment has finally been devised to remove them from their resting place.

Generations have come and gone without developing an efficient method and yet you can now go out alone and gather it all up. For the goldseeker, this is the age of opportunity.

Having checked the bar now you simply follow the river upstream. Ideally, you should dig a trench from one bank to the other with the dredge nozzle. Then, as you move up, the gravel you work fills in the trench behind you. Every trench you make affords the chance of unveiling a hiding place for gold.

You may come across a depression packed with sand and gravel which is sealing gold underneath. Even a granite bedrock, which is not prone to heavy wearing, has slight indentations which could be enough to hold gold if the sand is packed in hard above by rocks and the current is slow. Experienced dredge-operators who dive to the bottom of Australia's deep rivers are also familiar with the 'sticky' feel of what they refer to as a 'decomposed shale bottom'. This is often tacky enough to hold fine particles of gold. The dredge-operator should also know about the gold-retaining characteristics of dykes, although these are not common in the old water-courses of Australia where anything that was once sharp and jagged has been worn smooth by water. Many of Australia's underwater rock formations have been ground right down by the ageless flow of water but that does not preclude even the roundest of boulders and the shallowest of depressions from retaining gold. The rocks were sharp and jagged once, when they would have trapped gold, and although the angled edges that originally snared the metal have been worn down, many rock formations still hold their valuable catch.

Dykes are fingers of 'intrusive igneous rocks' in a softer type of bedrock. What that means is when the dyke was in its molten state it thrust itself into weaknesses in another, softer, rock and then hardened. The softer rock wears faster — in terms of thousands of years — than the dyke, so a 'tooth' of the dyke is left sticking up through the bedrock into the sand.

This tooth may be only 3 millimetres high but, if it is on an angle leaning against the downstream current, gold will be trapped and compacted into the corner between the side of the dyke and the surface of the bedrock. Even if the dyke is angled the other way, leaning with the direction of the current, it still has an ability to retain gold in much the same way as the riffles in the sluice box of the dredge. Gold particles sweep over the top of the tooth, and the eddy created by the current sweeps them back against the dyke in a similar action to the wash of a speedboat drawing in debris.

There are other areas of the river that the dredge-operator can investigate and which have much the same gold-stopping effect as the dyke. These are the spots where rocks or outcrops protrude from the bank part-way across the river. Many of the outcrops will have been worn back but they are nevertheless worth checking. If they are angled in an upstream direction they are a natural collector of rubbish on the surface and gold on the bottom. If the current is unusually slow, debris will be piled high, so a lot of messy clearing work with the spade will be needed before you can get the suction nozzle to the spots where gold may have collected. Alternatively, if an outcrop from the bank leans with the current, it is possible that gold will have swirled around the point of the outcrop and then tucked itself in behind it. The rule to remember is that if there is any protrusion from the bank into the river, investigate the point of the sharpest angle. The sharper the angle, the more easily gold will collect; it seems to prefer corners and angles. The same can be said

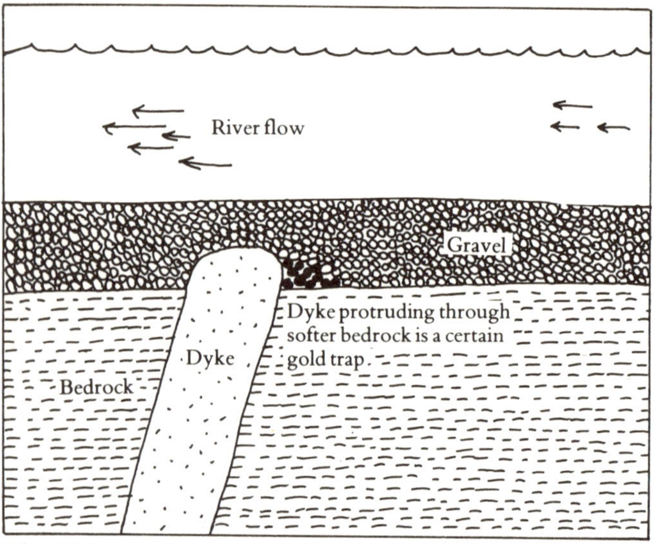


**The slightest obstruction will slow and perhaps trap gold as it moves with the river gravels. A dyke, a hard finger of rock, can act as a barrier.**

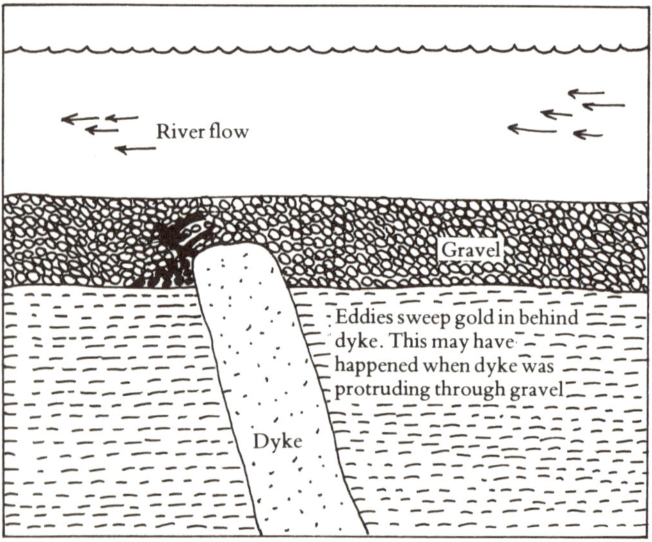


**Even dykes leaning with the flow of water can act as a gold trap by creating eddies.**

about the riffles in a plastic gold dish. Unless you are really careless, gold will stay close to them.

Crevices in the bedrock run in all directions and they should be investigated, particularly those that run across the flow of water out from the banks. Your chances are even better on the inside of a bend where the current runs slower; if you find a crevice here, give it thorough attention with the suction nozzle. Nor should cracks that run in the same direction as the current be ignored. Those that are very straight and shallow with smooth sides are not the best gold traps but they should still be checked. Possibly a stone has lodged in them and is acting as a blockade.

Sometimes even the most efficient dredge is incapable of extracting particles of gold that have become tightly lodged in crevices. Over a period of hundreds or thousands of years, the current and the movement of sand and gravel on the river bed has pressured and pushed tiny particles of gold to the very deepest slits and corners in the bedrock. The use of brute force is really the only way to get them out. If the water and the overburden are shallow enough, you may be able to reach down and prise the crevices open with a screwdriver or a trowel. If they are too deep you will need to dive down with a snorkel, but that is an aspect of dredging which is beyond the basic steps of prospecting outlined in this book.

The metal sluice box fitted to the modern dredge is a small, efficient version of the wooden trough used by earlier prospectors. Generally, the old sluice boxes were a series of troughs that slotted into each other on a downhill gradient from a hillside to the river. Stones, gravel and dirt were spaded in at the top and water carried them down through the troughs. Gold was trapped on the way by either riffles or perforations in the base of the trough, and the worthless material passed over the top to be deposited into the river. The principle of the sluice box is actually a copy of nature's work on the river bed, that is, gold travelling in water and becoming trapped by obstructions. Some sluice boxes were used in streams and were positioned so that 5 to 7 centimetres of water could flow through. The water was regulated by a gate at the top end of the troughs which could be raised or lowered.

Sluice boxes can be used to great effect to extract gold from elevated alluvial deposits, providing some means is found to get enough water flowing. If you invested in a small portable fire-fighting pump you would be able to wash down large amounts of gravel and earth from possible gold-bearing ground. Deposits of sticky clay should be kept out of the sluice box because it has a notorious tendency to gather gold, and as it is being washed through the sluice box it rolls into balls which will not break up. The best way to dissipate clay material is by puddling.

You can puddle in your panning dish or you can do it on a bigger scale by using a plastic bucket. In the gold-rush days prospectors used half a wine cask or a hollow log with both ends sealed so that it resembled a dug-out canoe. The puddling system is simple. Clay material is poured into the container and then covered with water. The contents are stirred until the mixture looks like a large volume of brown Windsor soup or mock turtle, depending on the area. A small amount of the surface is poured off and more water is added. The mixture is stirred again and more is poured off. The process is continued, with the liquid becoming thinner all the time, until only sand, gravel and gold remain. If enough running water is available the system can be short-cut by tilting the container and allowing the water to pour in over the top edge while the mixture flows out the lower side. When only sand and gravel remain at the bottom this can be panned or sluiced, depending on the size of the operation. Early prospectors who wanted to puddle large amounts of material would cut a round trough about a metre deep and 1 to 1½ metres wide with a big enough circumference to allow a horse attached to a beam to work around it. The other end of the beam was pivoted from the 'island' of ground inside the trough ring. As the horse walked it rotated the beam, which in turn dragged two triangular harrows through the material being separated. This sludge was run off from time to time and more water was added. There was little chance of losing gold at the bottom of the trench as it was lined with tightly-fitting slabs.

The metal cradle available in prospecting supply shops is a sophisticated version of the type used at the diggings in the 1850s. This is yet another way of separating gold from unwanted material, and the early prospectors found it to be a successful system. Dirt that needed washing was slushed through a square sieve; as the cradle was rocked on its stand by the operator shaking a fixed handle at the back, the material fell through the sieve on to a sloping board underneath it. As water poured through, the material fell from the sloping board, landing on a second sloping board where it continued its slide. It would then fall onto a third board. As it fell it left the gold behind on the boards. Cradles are very effective and fairly easy to make in any size. They were simple in the old days, too, and they had a hypnotic effect on the men who worked them.

John Sherer, who travelled out from England to seek his fortune on the goldfields in the early 1850s, wrote about prospectors who worked cradles:

All the operations and implements connected with the searching for gold are exceedingly simple, but the rough hairy men you see engaged in it here perform their duties with an earnestness scarcely to be imagined. They rock their cradles with all the seriousness of the tailor of Fife, who had four twins in his, and scarcely anything to put in their mouths; every faculty of their minds being concentrated upon 'the one thing needful' to enable them to realise the darling hobby of their existences, whatever that may be. Men who, all their lives, never knew what it was to have the smallest shade of gravity upon their brows, at once become metamorphosed into monks the moment they take the rocking stick in their hands. They become silent, solemn and scrutinising; fixing their eyes upon the dirt they are washing with miser-like intensity and altogether looking very unnatural in the sight of one who may be calmly surveying them. Alas that there should be so much care in seeking for that which 'maketh itself wings and flieth away, as an eagle toward heaven' . . . This is the process carried on from 'morn till dewy eve'; and unless it were diamonds or pearls, we know of no other sparkling pieces of matter on the face of the earth which could excite such a careful and curious anxiety in the face of man. Their gaze into the sieve is full of the deepest interest, and the half suppressed sigh which often immediately succeeds its conclusion tells of the disappointment which still meets the expectations of the gold seeker.

Where water was scarce prospectors devised a variety of cunning methods to separate gold from gravel. One of the more popular contraptions in earlier days was the dry blower, which came in several designs and could be mistaken today, perhaps, as a reject from Dr Who or from the film set of H. G. Wells' *Time Machine*. It stood as tall as a man, with funnels, bellows, supports and traps. It is best left to the early Western Australian prospector M. J. ('Mulga Mick') O'Reilly to describe it and the dry-panning process which followed its employment:

A dryblower is a bulky machine built with four legs, which must be of pliable wood. A square-built hopper is on top into which the wash-dirt is shovelled. As the dryblower is shaken, the fine wash percolates through the holes in the hopper on to a tray set on an angle of forty-five degrees. On this tray are twelve ripples, the bottom of the ripple-tray being finely perforated, to allow the blast of air from a bellows rigged underneath to blow the dust out of the sieved gravel that descends in a stream from the hopper above.

As the sieved and air-blown gravel slowly passes over the ripple-tray the gold is caught on the ripples. Large nuggets are sometimes found in the hopper tray. When all the ripples are full the tray is taken off and emptied into one of the two gold dishes. Standing sideways to the breeze, then, holding the filled dish high, one cants it gently so that its contents cascade into the other dish at one's feet. The breeze blows the dust and finer particles of light gravel out of the stream, but the gold and heavier stuff falls unerringly into the dish. Then you reverse the dishes and continue the process until satisfied that no clogging dust remains.

The half-filled dish is held knee high, its steel sides gripped firmly with both hands, shaken vigorously to make such gold as it may contain sink to the bottom. Then the dish is lightened by skimming off the top laying of dirt. The dish is again shaken and swirled and again lightened by skimming off another layer, and so on to the end when little remains. By working the dish with a rotary motion the little that is left is coaxed into one side of the dish. The dish is lifted close to the face and the residue is softly blown upon. Any gold in the dish will show up dull, bright and yellow.

Another contraption in dry areas was the shaker which rocked dirt through meshes of different sizes and trapped gold behind riffles. The shaker and the dry blowers have been replaced today by 'electro static concentrators.'

These appliances work on the same basic principle as the old dry blower but they have been developed beyond compare. The more expensive versions, about $1000, now supply warm air by means of an engine-driven blower assembly, a system that ensures that every particle of material to be processed is dried out enough to be separated. A steel screen holds back large waste rocks while the light waste dirt is blown away. Particles of gold are held by a static charge on the riffle tray.

One of the more intriguing new gold-recovery devices to appear on the Australia market is the goldspinner, whose American manufacturers claim it is 'super efficient in the recovery of gold and other precious metals'. In simple terms, it is three gold-pans in one. Material saved from the first pan, or wheel, goes into the second for further processing and what is saved at this stage is transferred to the third wheel for yet another separation process.

Like any electronic device designed to recover gold, the three-wheeled goldspinner is not cheap; it sells for $1285. However, for the amateur it is claimed to be the fastest means of obtaining professional results in gold recovery. The prospector can also check locations much more quickly than with the traditional panning method. When set up beside a river the wheels, which have grooves in their base, are at an angle behind one another, each angle being slightly different. Material collected for separation is poured into the first wheel, and water then automatically washes the light waste out over the rim. The remaining concentrate passes through the centre of the wheel to the second wheel where it is separated further. By the time the material has been processed in the third wheel, gold in the initial pile of gravel and dirt will be apparent.

With so much equipment at his disposal for different locations, the modern goldseeker has every chance of putting back into his wallet that which he has spent. Of course, you don't need to spend *anything*. Some old-timers have been known to extract gold from rivers with a frying-pan. But that's another story . . .

# 12 How to Sell Gold

How much gold is worth depends not only on the quantity but on weight, shape, colour and texture. Flake gold weighing 186 grams (6 ounces) and a nugget weighing 186 grams may share the same weight, but they certainly don't share the same value. You can expect to sell the nugget for twice the amount of the market value of gold.

On the other hand, if you have only flake gold weighing 186 grams don't multiply each gram by the current value of gold, and rush out and invest in a fur coat. Your 186 grams may be worth only three-quarters (or less) of the value of gold on the open market, and here is the reason.

Twice a day, at 10.30 a.m. and 3 p.m. in a modern building in St Swithin's Lane, London, five soberly dressed young men sit down at separate tables, on each of which are placed a telephone and a small British flag. The young men are representatives of five leading houses in the gold trade and the man who chairs the meeting, Robert Guy, head of Rothschild's gold and foreign exchange operations, has to find a consensus price that will satisfy all five firms. The price established at Rothschild's is based not on 18-carat or 22-carat gold but on pure 24-carat gold. So when they agree that gold should be fixed at, say, $620 an ounce, they are talking of the finest quality.

Now, the flakey gold from the rivers of Australia or the small nugget dug from around the flats isn't '1000 fine'. At the very best it is, generally speaking, around 95 per cent gold, that is, .950 fine. So, depending on the current price, you are already down on your initial expectations. There are also charges. Some refiners, to whom most particle gold is sold, will not take less than 311 grams (10 ounces), while others will happily accept 10 grams from the weekend prospector.

One Melbourne refiner says: 'Our refinery charges are $54, made up of $24 for an assay and $30 for the refining. As at February 1980, a prospector who comes to us with 2.7 grams of gold breaks even because that, on average, is what his gold is worth and that's what we charge for handling it. Anything more than 2.7 grams and he starts to make money. It isn't really worth his while coming in with less than 10 grams. All he need do is just wait until he's collected that amount — it's not much. He gets his money eight days after he brings the gold in.'

There may be additional charges at refineries for the separation of one or two metals from gold. To establish exact figures you need to take your sample to the refinery.

The Mines Department staff are always concerned about the efforts of the small-time prospector and will be more than willing to give you extra advice on samples. They will also carry out an assay for you and, although this is free at the time of writing, charges may be introduced.

While it is still very profitable to gather up reasonable amounts of the small stuff, the real money lies in nuggets, which says much for owning a metal-detector. What you *don't* do when you dig up a 311 gram (10 ounce) specimen is run straight to the refiner's with it. A very frank Victorian refiner advised: 'If you've got a nugget, don't try to sell it to me. You'll get twice the market value if you sell it to a collector.'

Some nuggets will be worth more than twice the market value. Much depends on their shape and the richness of their colour. A nugget with quartz showing through will be worth even more; more, that is, than a nugget of solid gold of the same weight! A nugget with quartz that has been coloured by traces of iron is rare and so can your price be. If the quartz is a soft pink its value is very high. The attractiveness of nuggets is that if you had one of 186 grams (6 ounces) worth about $6000 and the price of gold was to take a temporary tumble, your nugget would remain at its nugget value.

If you have a nice nugget and you really do want to part with it, you need a buyer who will offer you the right price. Such a buyer is becoming a little more difficult to find because a man who has already bought six each weighing 311 grams (10 ounces) isn't likely to want another seven or eight unless they are shaped like Snow White and the Seven Dwarfs. You can make all the obvious telephone calls to the Mines Departments and to the refineries, asking if they know of a collector who is buying nuggets. If that fails you can always advertise your specimen in the newspapers.

The beautiful thing about gold is that if you need cash you will always be able to get it, even if you end up selling your gold to one of the scrap-dealers in precious metal who are to be found in most big cities and who will give you a reasonable price.

Remember, there's always someone who will buy your gold if you feel you really must part with it.

# 13 Where to Go: State by State

**Victoria**

Miner's Rights are available from most prospecting supply shops as well as from the Department of Minerals and Energy, Russell Street, Melbourne. The fee is $2.50 and the certificate, which is in force for a year, entitles a prospector to use a metal-detector or a pan on Crown Land. He may use a pick and a shovel but he cannot use machinery or explosives and he cannot build dams. Metal-detectors are not classified as machinery.

A prospector can peg a claim for gold on Crown Land with a maximum area of 1 hectare and after the claim has been issued he can use machinery and explosives in working it.

The holder of a Miner's Right is also entitled to peg a claim on private land which can then be prospected with machinery once the claim has been issued and the question of compensation is set. Pegging out a claim involves the placing of marker posts on the corners of the land. Registration of a claim costs $2.50 and it has to be worked for eight hours or more in any period of fourteen days.

Dredging is permitted only on sections of the Goulburn River and its tributaries upstream of Jamieson, after the issue of a Search Licence (fee $12). An application must be made to the department to dredge in other rivers.

**Goldfields** Alma, near Maryborough, leads found from 2 m to 9 m; Ararat, 205 km north west of Melbourne, alluvium and basalt-covered leads; Ballarat, 110 km north-west of Melbourne, rich alluvial deposits and deep leads; Beechworth, 32 km south-west of Wodonga, alluvial gold from river areas; Bendigo, 138 km north-west of Melbourne, quartz veins and rich alluvial gullies; Blackwood, 76 km north-west of Melbourne, shallow alluvial deposits; Blakeville, 75 km north-west of Melbourne, rich alluvial deposits; Blue Mountain, 32 km south-east of Daylesford, alluvial deposits in gullies and creeks; Bright, 115 km south of Wodonga, reefs and alluvial deposits; Buninyong South, 12 km south of Ballarat, rich leads; Cassilis–Bethanga, 50 km east of Bright, small

# Victoria

reefs; Castlemaine, 120 km north-west of Melbourne, rich ground from 1 m to 10 m in depth; Chiltern and Rutherglen, 42 km west of Wondonga, deep leads; Clunes, 30 km north of Ballarat, deep leads; Creswick, 15 km north of Ballarat, shallow alluvial deposits; Dart River, 80 km south-east of Wodonga, alluvial deposits in rivers and creeks; Daylesford, 110 km north-west of Melbourne, rich alluvial deposits; Dunolly, 23 km north of Maryborough, rich alluvial deposits; Egerton, 26 km east of Ballarat, rich alluvial deposits; Elaine, 42 km south-east of Ballarat, rich alluvial deposits; Fryers' Creek, 18 km south-east of Castlemaine, heavily worked area, alluvial deposits up to 15 m; Ghin Ghin, on the Goulburn River, 10 km south of Yea, alluvial deposits, quartz reefs; Gordon, 20 km east of Ballarat, alluvial deposits; Harrietville, 25 km south of Bright, reefs, some alluvial deposits along Ovens River; Heathcote, alluvial gold from Mount Ida Creek, 110 km north-west of Melbourne, and from its tributaries; Hepburn, on the southern sources of the River Loddon, alluvial deposits in streams and gullies; Kingower, 58 km north of Maryborough, rich source of alluvial nuggets; Malmsbury, 24 km south-east of Castlemaine, some alluvial deposits in gullies and creeks; Maryborough, 56 km north of Ballarat, alluvial deposits in gullies and creeks; McIntyre's, 48 km north of Maryborough, rich producer of nuggets; Mitta Mitta, 105 km south-east of Wodonga, rich reefs; Moliagul, 40 km north of Maryborough, shallow alluvial ground where the Welcome Stranger was found and a source for other nuggets; Moorabool, 6 km north-west of Geelong, alluvial gold in creeks and gullies; Newstead, 16 km south-west of Castlemaine, alluvial deposits; Poseidon, 43 km north-east of Maryborough, shallow nuggets; Pyrenees Belt, 265 km north-west of Melbourne, alluvial gold in Moonambel, Lambing Flat and Redbank; Rheola, 52 km north of Maryborough, rich in alluvial nuggets; Rushworth, 145 km north of Melbourne, alluvial gold; St Arnaud, 265 km north-west of Melbourne, alluvial gold; Steiglitz–Maldon belt, taking in Steiglitz, 30 km north-west of Geelong, Yandoit, Egerton and Daylesford, to Maldon, 20 km north-west of Castlemaine, yields of alluvial deposits up to 10 m depth; Taradale, 18 km south-east of Castlemaine, rich ground up to 10 m; Tarnagulla, 40 km north-east of Maryborough, deep reef deposits; Yackandandah, 295 km north-east of Melbourne, alluvial gold around Nine Mile Creek and Woolshed Valley; Yarra River, from the Baw-Baw Mountains to Port Phillip Bay, gold-bearing creeks and gullies; Yea–Alexandra, approx. 120 km north-east of Melbourne, alluvial deposits in the Acheron River and streams; Warrandyte, 27 km east of Melbourne, small reefs; Walhalla–Gaffney's Creek, 130 km east of Melbourne, deep reefs, alluvial gold in rivers; Wedderburn, 75 km north of Maryborough, reefs and nuggets.

**Nuggets** A large number of nuggets has been found at shallow depths in many parts of Victoria, most coming from the goldfields to the north-west of Melbourne. A full list of nuggets found, including their weight, location and depth, is available from the Department of Minerals and Energy. The areas that have produced the highest number of shallow nuggets (that could have been found with a metal-detector) are: Rheola (originally known as the Berlin Goldfield), 38 shallow nuggets; Wedderburn, 24; Dunolly, 21; Moliagul, 20; Kingower, 18; Poseidon, 10; Bendigo, 10; Tarnagulla, 7; Buninyong South, 7; Hepburn, 6; Queenstown, 6; Blackwood, 5; Heathcote, 5.

**New South Wales**

A Fossicking Licence entitles the holder to search for gold in prescribed Fossicking Areas on Crown Land. Personal licences cost $2 and a family licence is $4. They are valid for one year. Licences are not required on private land, but the consent of the owner must be given. Fossicking is allowed only in the fifty-four areas set aside throughout the state and outside these areas gold, being a royal metal, remains the property of the Crown. Some of the fossicking areas are on private land but most are on Crown

Land. Detailed maps of the areas are available from the Titles Branch of the Department of Mineral Resources and Development, C.A.G.A. Centre, Bent Street, Sydney, and at country centres.

The fossicker is entitled to 10 kilograms of gold — about a 9 litre plastic bucket full — in any 48-hour period. (The same rule applies for rocks or other stones.) Gold diggings can be to a depth of 2 metres. Anyone who is engaged commercially as a jeweller, mineral collector, lapidary or dealer in or seller of gemstones, semi-precious stones or rocks, is not allowed to operate in a Fossicking Area. Metal-detectors are allowed but other powered machinery is not unless a prospector has title to a claim. Fees for claim registration are $20 and a nominal amount of $50 is usually required as a guarantee that the claim-holder will repair damage caused by his work.

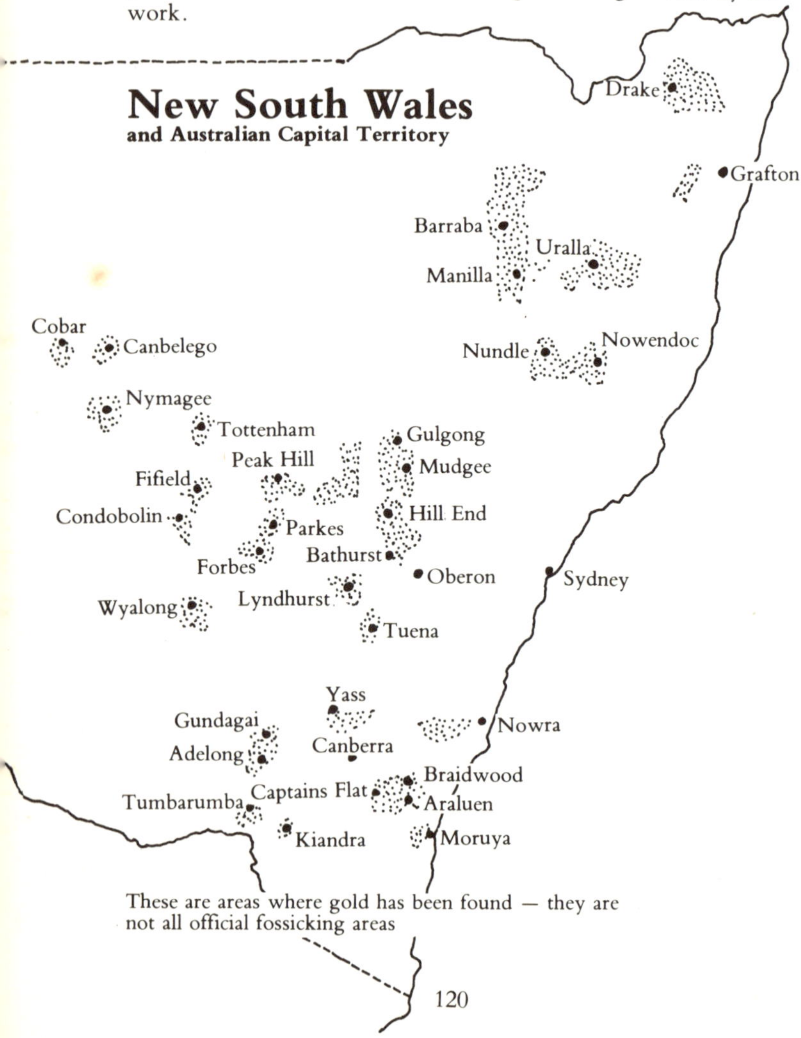

These are areas where gold has been found — they are not all official fossicking areas

Prospectors who wish to dredge in New South Wales should make an application to the Department of Mineral Resources and Development.

**Fossicking** Areas you are allowed to go fossicking in can be found at:

Parish of Tambaroora, County of Wellington, at Tambaroora; Parishes of Biala and Winduella, County of King, between Grabben Gullen and Bevendale; Stuart town, between Orange and Dubbo; Crown Mountain–Sara River, about 45 km south-east of Glen Innes; Kookabookra, 10 km east of Crown Mountain; Parish of Uralla, County of Sandon, about 3 km west of Uralla; County of Nundle, 3 km east of Nundle; Parish of Bombar, County of Georgiana, at Abercrombie Caves, between Abercrombie and Trunkey Creek; Ophir, about 20 km north-east of Orange; Parish of Wiagdon and Sofala, County of Roxburgh, at Wattle Flat and at Sofala; Parish of Young, County of Monteagle, at Young; Parish of Windeyer, County of Wellington, at Windeyer; Parish of Adderley, County of Westmoreland, between Black Springs and Essington; County of Géorgiana, Abercrombie River, near village of Abercrombie.

Detailed maps of these areas can be obtained from the Department of Mineral Resources and Development.

**Nuggets** 1851, Meroo Creek, River Turon, near Mudgee, 39 kg (1272 oz); 1851, Meroo Creek, 11 kg (364 oz); 1852, Meroo Creek, 4882 g (157 oz) and 2208 g (71 oz); 1858, Burrandong, near Orange, 4000 g (1286 oz); 1860, Kiandra, Snowy River, 12 kg (400 oz); 1861, New Chum Hill, Kiandra, 6220 g (200 oz); 1874, about 6 km from Parkes, 217 g (7 oz), 777 g (25 oz), 1088 g (35 oz) and 1150 g (37 oz); 1874, Woods Flat, near Cowra, 2021 g (65 oz); 1876, near Gulgong, 1990 g (64 oz); 1879, Nundle Gold Field, 1337 g (43 oz); 1879, Braidwood district, 995 g (32 oz); 1880, Temora, 1834 g (59 oz), 1959 g (63 oz), 1990 g (64 oz), 2114 g (68 oz), 2239 g (72 oz) and 2363 g (76 oz); 1880, Nerrigundah, at the foot of Mount Dromedary, 404 g (13 oz); 1882, Temora, 4758 g (153 oz); 1882, Cadia, near Orange, 2177 g (70 oz); 1882, gully leading into the Upper Meroo, 901 g (29 oz); 1884, Gulgong area, 12 nuggets ranging from 93–1212 g (3–39 oz); 1884, between Upper and Lower Temora, 21 nuggets ranging from 155–5442 g (5–175 oz); 1884, Mudgee mining district, eight nuggets ranging from 155–1212 g (5–39 oz); 1885, Cadia, 653 g (21 oz); 1887, Maintland Bar, Mudgee District, 10 kg (350 oz).

### Western Australia

A prospector wishing to search for gold must first buy a Miner's Right from the Mines Department. Fee: 50 cents. This entitles him

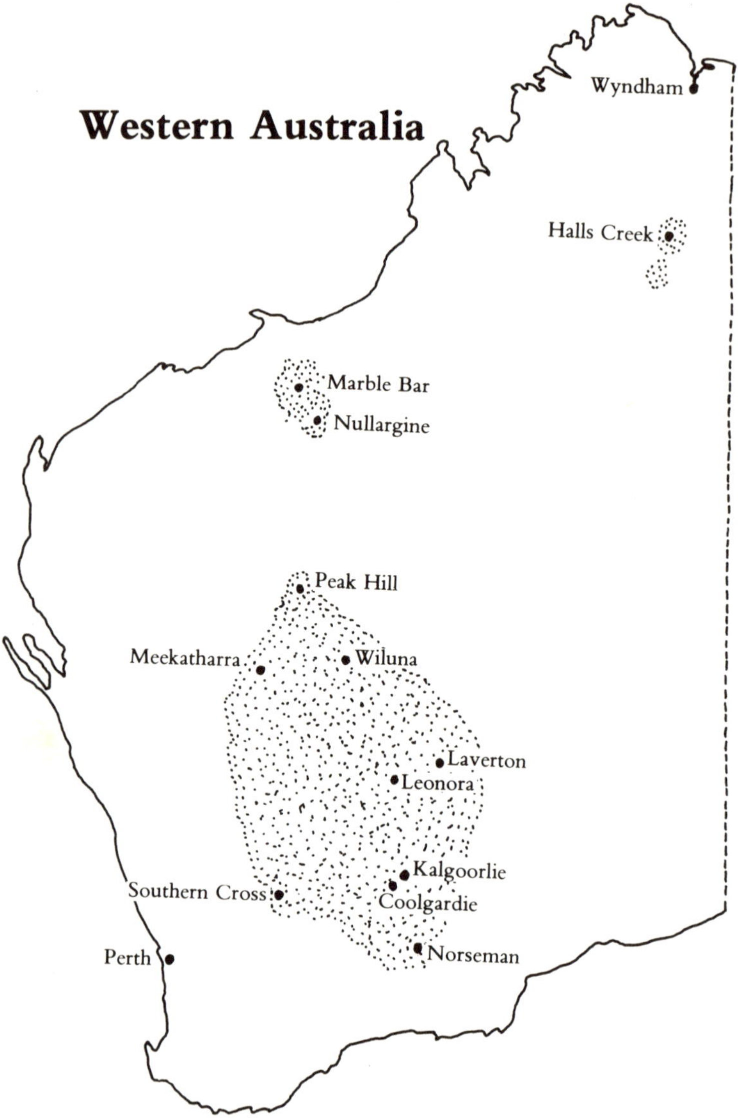


to peg, apply for and in due course hold most types of mining tenements. A prospecting area within a proclaimed goldfield can be as much as 9.7124 hectares, and if it is outside a proclaimed field the allowance is 19.4249 hectares. Dredging is allowed and the fees are 50 cents for .4047 hectares a year, plus a registration fee of 50 cents.

**Goldfields** In the early part of the century Western Australia had sixteen goldfields operating. These were: Ashburton, Coolgardie, Dundas, East Murchison, Gascoyne, Kimberley, Mt Margaret,

Murchison, North Coolgardie, North East Coolgardie, Peak Hill, Phillips River, Pilbara, West Pilbara, Yalgoo, Yilgarn.

**Nuggets** For the assistance of metal-detector owners, here is a list of nuggets reported from 1888 to July 1979, in chronological order of discovery:

1888, Pilbara Creek, 3910 g (127 oz); 1890, Coongan River, 10 kg (333 oz); 1891, Nannine, 12 kg (400 oz); 1891, Nannine, 3110 g (100 oz); 1892, Cue, 3110 g (100 oz); 1892, Lake Austin, 2799 g (90 oz); 1892, Pilbara Creek, 5131 g (165 oz); 1893, Top Camp, 2705 g (87 oz); 1893, Top Well, Murchison, 9268 g (298 oz); 1894, Londonderry, 26 kg (850 oz) — from reef; 1894, Wealth of Nations, Coolgardie, specimens from reef cap, 24 kg (800 oz); 1895, Black Flag, 12–15 kg (400–500 oz); 1896, Croydon, 4976 g (160 oz); 1897, Wiluna, 14 kg (463 oz); 1898, Kurnalpi, 3389 g (109 oz); 1898, Kanowna, 3203 g (103 oz); 1898, Bulong, 2488 g (80 oz); 1898, Bulong, 2332 g (75 oz); 1898, Bulong, 1648 g (53 oz); 1899, Sharks Gully, 15 kg (487 oz); 1899, Sharks Gully, 12 kg (372 oz); 1899, Kurnalpi, 5225 g (168 oz); 1899, Mt Monger, 4198 g (135 oz); 1896, Mt Mortimer, 1866 g (60 oz); 1899, Hogans Find, 3887 g (125 oz); 1899, Mt Monger, 3172 g (102 oz); 1899, Hogans Find, 3887 g (64 oz); 1899, Hogans Find, 1555 g (50 oz); 1900, Kurnalpi, 6127 g (197 oz); 1901, Coolgardie, 4510 g (145 oz); 1901, Black Hills, 1617 g (52 oz); 1901, Kurnalpi, 1586 g (51 oz); 1901, Black Hills, 1555 g (50 oz); 1901, Coolgardie, 1555 g (50 oz); 1902, Friendly Creek, 5131 g (165 oz); 1903, Shark Gully, 6500 g (209 oz); 1912, Kurnalpi, 7153 g (230 oz); 1912, Kurnalpi, 1399 g (45 oz); 1912, Ora Banda, 1337 g (43 oz); 1912, Paynes Find, 3421 g (110 oz); 1915, Ora Banda, 4820 g (155 oz); 1931, Larkinville, 35 kg (1135 oz); 1931, Larkinville, 3110 g (100 oz); 1934, Gullewa, 5131 g (165 oz) — 40 m below surface; 1934, Block 48, Hampton Plains, 1181 g (38 oz); 1935, Kalgoorlie, 497 g (16 oz); 1935, Siberia, 1959 g (63 oz); 1935, Coolgardie, 466 g (15 oz); 1935, Coolgardie, 529 g (17 oz); 1937, Boulder, 1461 g (47 oz); 1938, Ora Banda, 373 g (12 oz); 1938, Dunnesville, 559 g (18 oz); 1939, Norseman, 3172 g (102 oz); 1946, Cutters Luck, 1492 g (48 oz); 1948, Bonnievale, 404 g (13 oz); 1973, Mt Magnet, 1461 g (47 oz); 1978, Kalgoorlie, 839 g (27 oz); July 1979, Western Shaw, Pilbara, recorded value $17 000.

**Queensland**

If you are casually fossicking for gold with a metal-detector you don't need a Miner's Right. If you intend to do serious prospecting you must obtain a Miner's Right. The Queensland Department of Mines points out that the laws relating to the issue of Miner's Rights have been tightened owing to the 'numerous complaints that have been received' regarding wilful destruction of other people's property which in some cases amounted to 'criminal offences'. A written application on a prescribed form must be submitted and proof of identity of the applicant may be required. A Miner's Right costs $5 a year and entitles the holder to enter Crown Land for marking out an area that is to be mined or to occupy land for 'hand mining purposes'. This is defined as using only picks, shovels, hammers, sieves and similar tools which are used manually. As dredges are new in Australian gold-prospecting, Queensland, like most other states, has made no special provisions for dredge-operators. Specific enquiries should be made with the Department of Mines, Mineral House, 41 George Street, Brisbane.

Alluvial deposits have been found in a wide belt stretching from the New South Wales border to the tip of Cape York Peninsula. Almost all the state's goldfields have yielded gold in alluvial deposits.

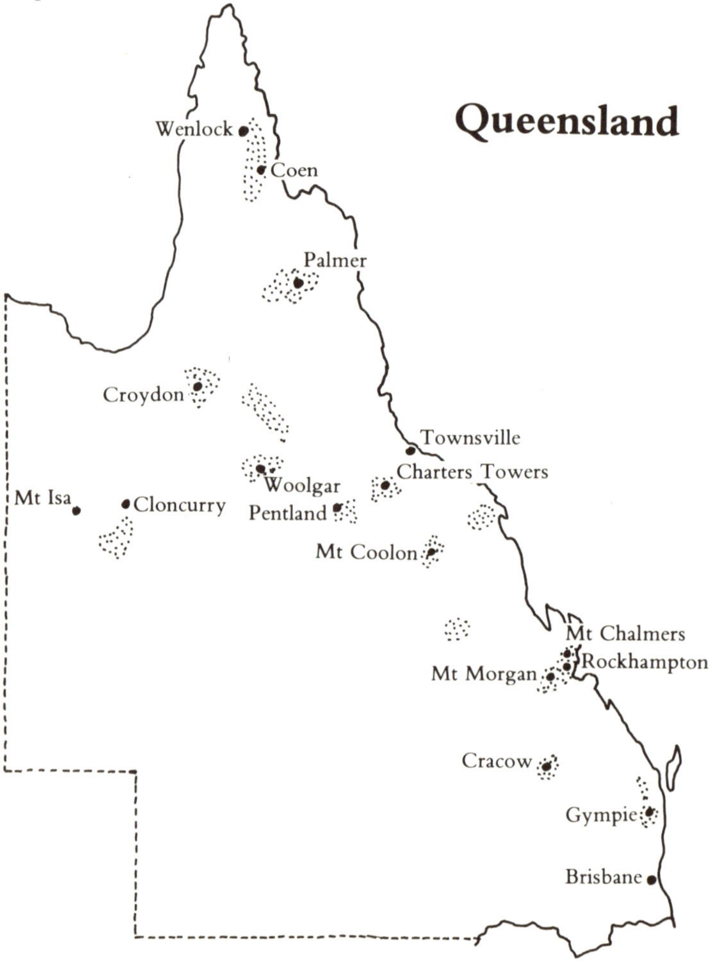

**Principal goldfield areas** Brisbane area, Bowen, Cairns, Charters Towers, Clermont, Cloncurry, Cooktown, Croydon, Einasleigh–Etheridge, Gladstone, Mackay, Maryborough–Gympie, Palmer, Ravenswood, Rockhampton. There are many goldfields scattered around each of these regions. The Department of Mines can give specific details of the areas you are interested in.

**Nuggets** There are records of at least fifteen Queensland nuggets weighing more than 3110 g (100 oz) as well as many smaller ones. Nugget discoveries include the following: 1868, Gympie Creek, 30 kg (975 oz); 1868, Sailor's Gully, Deep Creek, 25 kg (804 oz); 1868, Mt Wheeler, near Rockhampton, 8023 g (258 oz); early 1870s, Top Camp, Cloncurry, 10 kg (336 oz) as well as 2736 g (88 oz) from Two Mile diggings and others of 528 g (17 oz) and 311 g (10 oz) from Top Camp; early 1880s, Nuggety and Moonlight Gullies, 217–2488 g (7–80 oz); 1896, Green Hills, west of Forsayth, 60 nuggets up to 5598 g (180 oz); September 1903–August 1904, Dee River, near Mt Morgan, nuggets totalling 135 kg (4350 oz), the largest weighing 5660 g (182 oz); 1913, Plutoville, 3763 g (121 oz); 1920, near Mt Leyshon, 1447 g (143 oz). Other undated discoveries include nuggets of 2799 g (90 oz) at Ebagoola; 1741 g (56 oz) at Langmorn; three weighing 1492 g (48 oz), 933 g (30 oz) and 933 g (30 oz) at Clermont; various sizes up to 2021 g (65 oz) at Talgai; and one of 528 g (17 oz) at Cape River.

## South Australia

The weekend prospector is classified as a fossicker. Fossicking is defined as 'the gathering of minerals as a recreation without any intention to sell the minerals or use them for a commercial or industrial purpose'. The activities of the fossicker are not governed by the provisions of the Mining Act, although fossickers are urged to respect the rights of property-owners and ask permission to enter private land. A prospector who intends to seriously gather minerals should obtain a Miner's Right, which gives him the authority to peg out a claim. An application form has to be lodged with the Department of Mines and Energy, 191 Greenhill Road, Parkside, South Australia, and a fee of $10 must be paid. The certificate is valid for three years. Prospectors wishing to dredge should discuss their plans with the department as each case is judged on its merit before permission can be granted.

**Principal alluvial goldfields** Algebuckino, Angepena, Barossa, Bismarck, Boolooroo, Chapmans Gully, Clare, Daws, Deloraine, Golden Belt, Golden Gate, Jupiter Creek, Lady Alice, Mintaro, Moppa, Mt Monster, Mt Ogilvie, Patawurta, Stirling, Taltabooka, Teetulpa, Ulooloo, Watts Gully.

**Largest nuggets reported** Teetulpa, 2488 g (80 oz), 901 g (29 oz), 435 g (14 oz), 404 g (13 oz), 342 g (11 oz), 311 g (10 oz); Scott's Diggings (Cromer), 1088 g (35 oz), 1492 g (48 oz), 933 g (30 oz); Forest Range, 1492 g (48 oz), 311 g (10 oz); Watts Gully (Mt Pleasant), 684 g (22 oz), 1492 g (48 oz), 279 g (9 oz); Blackwood Gully Diggings, 217 g (7 oz); Echunga, 373 g (12 oz), 217 g (7 oz).

**Northern Territory**
A casual gold-fossicker with a metal-detector or a panning dish does not need a Miner's Right. However, if he is seriously looking for gold he must purchase a Miner's Right, which is 50 cents a year. Anyone who finds an area he may want to extract gold from can register a 'prospecting area' at $1 a year — from which he is allowed to take up to 50 tonnes of material for testing. On goldfields, these areas can cover about 8 hectares; off goldfields they can be as much as 16 hectares. New mining regulations, which are expected to be introduced in 1981, will probably include 'Fossicking Areas' as in New South Wales.

Many of the Northern Territory's gold deposits lie in small goldfields about 195 kilometres south-east of Darwin. They are found within a radius of 75 kilometres from Pine Creek and the gold is in quartz reefs, quartz ironstone and alluvial deposits.

**Goldfields** Gold lies in the following areas: Brocks Creek, Daly River District, MacDonnell Ranges, Maude Creek, Pine Creek Field, Tanami District, Tennant Creek, and Union Reefs Field.

**Nuggets** More than 100 nuggets have been reported found within a 100 km radius of Pine Creek. They ranged from 62–2177 g (2–70 oz), more than 50 weighing over 622 g (20 oz).

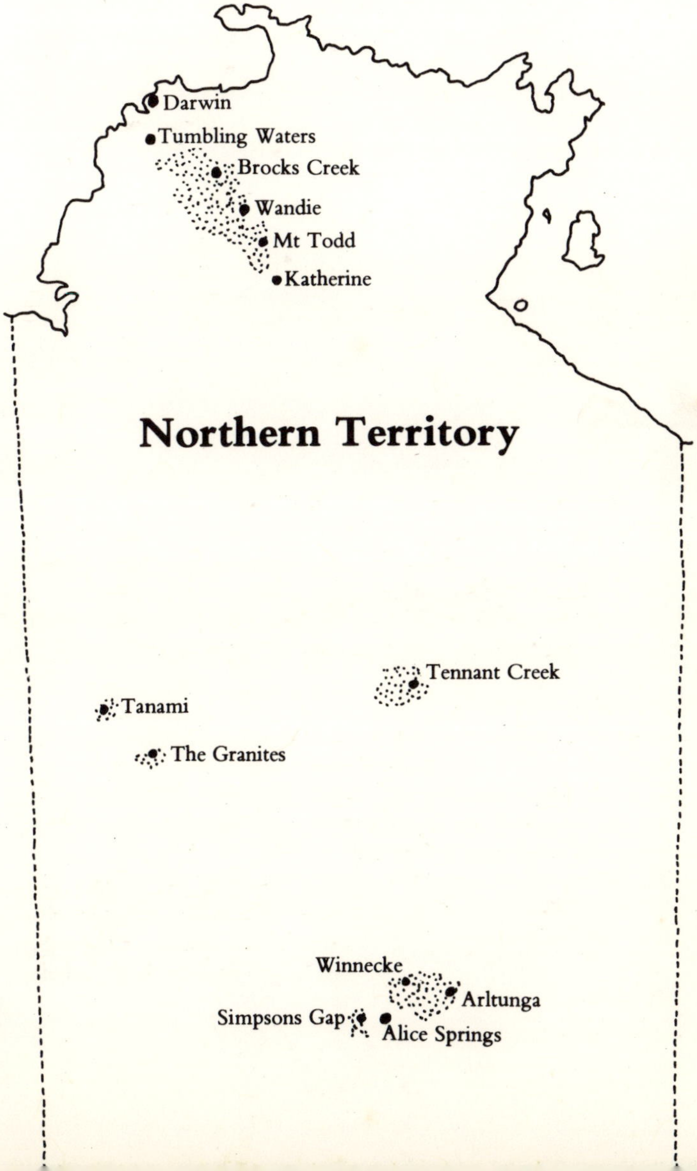

**Tasmania**

Tasmania has not produced large amounts of gold and the Department of Mines has made no special provisions dealing with amateur gold-prospectors in the state. Prospecting cannot be undertaken either on Crown Land or private land unless it is in accordance with the state mining legislation. Serious goldseekers can obtain a Prospector's Licence for $10 a year and this authorises the holder to take possesssion of 25 hectares of unoccupied Crown Land for prospecting purposes. Limited quantities of samples may be removed for testing and assay. A Miner's Right, at $5 a year, entitles the holder to mark out and take possession of a quarter of a hectare of unoccupied Crown Land for the extraction of minerals.

The most important sources of gold have been the Beaconsfield, Lefroy and Mathinna goldfields in the north-east region, although gold has also been found on the west coast near Queenstown, on the north coast at Burnie and at Cygnet, south-west of Hobart.

Museum records reveal that the only two large nuggets found were in 1883 in the Rocky River. They weighed 7557 g (243 oz) and 4447 g (143 oz).

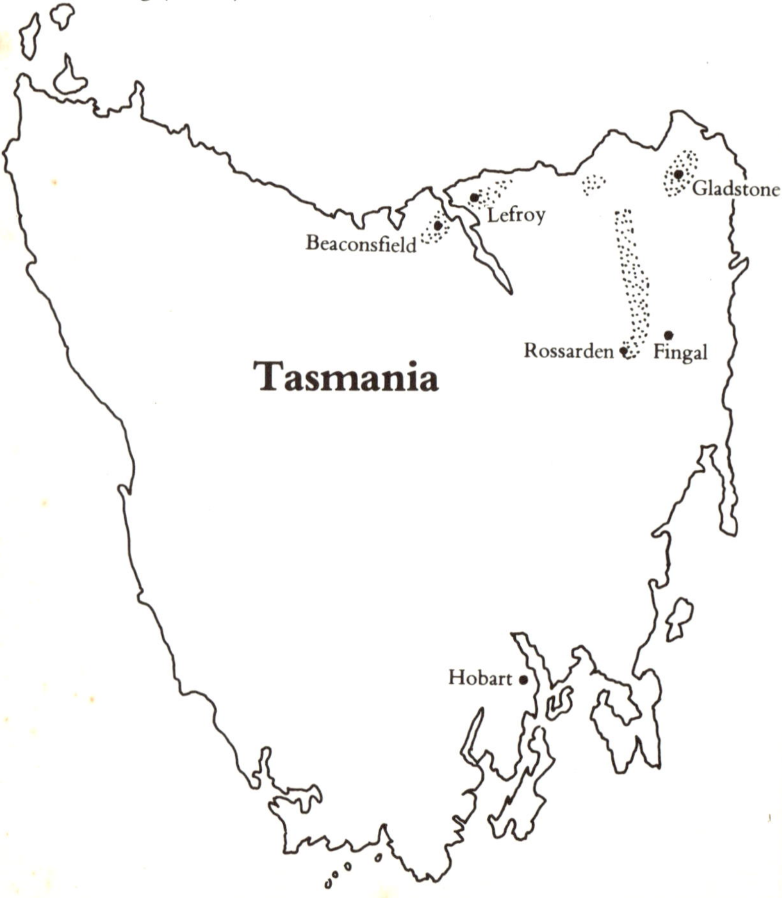